Building in Africa – Cape to Cairo in 150 days: Experiences of an Engineer

Mike Schlaich

Building in Africa – Cape to Cairo in 150 days: Experiences of an Engineer

Mike Schlaich

DOM publishers

Contents

Where an Engineering Safari can lead

When we Europeans think of Africa as a geographical area, we can probably list all 54 countries. But when it comes to naming the capitals, many of us find it more difficult. Most people are surprised to learn that more than 2,000 languages are actively spoken on the continent. This is because many African countries still use the official languages of their former colonial powers in economic life – primarily English, French, and Portuguese. For most countries, emancipation from foreign rule began in 1960, the so-called African Year. Seventeen colonies and one trust territory were granted independence. Until then, the cultural areas between the Mediterranean, Atlantic, and Indian Oceans had been blank spaces on the map of European strategists for decades, in some cases centuries.

When the Congo Conference took place at the turn of 1884/1885 in the Reich Chancellor's Palace in Berlin at the invitation of the German Empire, Africa was still an unknown continent for Europeans. Influential European monarchies – initially Portugal, later also the Netherlands and Kurbrandenburg as well as Sweden and Denmark – had built fortifications along the western coasts from the late fifteenth century onwards to provide their fleets with a secure stopover on their sea route to India and to promote trade in gold, ivory, and slaves. On the east coast, Arabs and Indians had long been engaged in lively trade and had shaped the cultural area there. However, with a few exceptions, contact was limited to the coastal regions. Until well into the mid-nineteenth century, only a few expeditions ventured to explore the interior. The 'hinterland' – a German term that entered English colonial language – remained unexplored by the curious and at the same time imperialistic Europeans. In terms of construction, it was still too complicated to expand the existing paths of the indigenous population to meet the needs of the intruders. Economically, it was of no interest to the colonial bases on the coast, as they conducted their local trade with Africans who knew their homeland much better and had easy access to the desired resources. In the northern hemisphere, the African 'hinterland' was stylised by those returning home as a mythical place: a jungle thicket inhabited by exotic animals and people with strange customs.

Let's be honest: These images are still anchored in our minds today. The African landscape is a fascinating mosaic of cultures, traditions, and remarkable natural beauty. Just 150 years ago, the people of the continent lived in harmony with nature. Industrialisation, urbanisation, and the infrastructure required for this simply did not exist. But that has changed within a few generations. Behind this often romanticised 'facade of the primitive' lies an impressive development: civil engineering, which is the driving force behind the creation of infrastructure on a rapidly changing continent. Precisely because Africa has these vast natural spaces, engineering structures are of existential importance to the local people. The tradition of this type of construction is relatively young. Only in North Africa did the Romans and other Mediterranean cultures leave behind viaducts, bridges, and port facilities.

Knowledge of these structures did not spread south across the Sahara. The few exceptions of technically complex structures in sub-Saharan Africa remain a mystery to construction researchers to this day.

Viewing Africa through the eyes of a civil engineer opens up entirely new perspectives. That is what Mike Schlaich does for us. With a keen eye for the realities and possibilities of Africa, the globally active engineer opens our eyes to the vital connections represented by bridges, dams, and railway lines. These infrastructures are much more than just technology. They are the lifelines of civilisation, enabling people to realise their dreams and ambitions. Schlaich does not hide the fact that many engineering structures were built 100 years ago to exploit colonies. While we in Europe often have only limited insight into the complex, dynamic developments in Africa, the professor, who teaches at the Technische Universität Berlin, explains in an unpretentious way how the continent's urban population will change dramatically in the coming years. Africa's future is in flux, with signs pointing to growth – by 2050 an additional 1.3 billion people will be living in Africa. That is three times the current population of the EU.

Whenever we look at Africa, we often see only the political conflicts and demographic challenges. But Mike Schlaich's photographs from his trip and his extensively researched diagrams show us the other side: the curiosity and thirst for knowledge of the people, their initiatives to improve their living conditions, and the diversity of their cultural identity. In his diary, he shares very personal experiences and profound reflections that show us how important it is to view Africa not only as a recipient of aid, but as an active player on the global stage whose voice must be heard and respected.

Mike Schlaich calls his 150-day journey from Cape Town to Cairo an 'engineering safari': a neologism that could hardly describe Europe and Africa better in this context. On the one hand we have the Latin *ingeniator*, one who solves a technical problem, and on the other we have *safari*, the Swahili word for journey. On his engineering safari, Mike Schlaich did not set out to solve a technical problem. But he gives us insights into contemporary planning and construction on the continent and encourages us to think about the future of cooperation and Africa's new role in global society. So, let's dive into an African history of infrastructure that is more than the sum of its technical structures.

Philipp Meuser,
architect and publisher

Foreword and Acknowledgements

The 'engineering safari' described in this book was the journey of my life and an incredibly insightful research semester on construction in Africa. My friend Massimo Canossa, a tropical agronomist, and I, a civil engineer, chose the route through 12 of Africa's 54 countries together during the Covid summer of 2022, so that it would still be possible to travel through all the countries. My plan was to get an idea of civil engineering on this continent. The impressions and insights gained from this road trip and my subsequent research are, of course, not enough to make me an expert on construction in Africa. However, outlines are emerging, and because little has been published on this topic so far, I have taken the liberty of reporting on my journey in the following pages.

Writing about it was more exhausting than the 150-day drive from Cape Town to Cairo itself. I live in Germany, a constitutional state with a social market economy, a democratic system that I value. I have realised that I cannot simply transfer my view from affluent Western society to Africa. Different standards often apply on this continent, and one is confronted with leaps in scale that are unknown in my country.

For example, the term 'infrastructure' – the construction of cities, roads, railways, bridges, ports, airports, and power stations – no longer has only positive connotations in Germany. In Africa, it usually does, because so much is still lacking. The damage caused by colonialism in Africa cannot be undone and I want to take a pragmatic view of things that I can no longer change. What can be done? What is feasible? I do not want to stay out of it – I embarked on this journey to experience (*erfahren*) the continent in the truest sense of the word. I have tried to gather facts and at the same time not hide my opinion. As an 'old white man', this makes me vulnerable. Cultural appropriation and neo-colonialist know-it-all attitudes may have crept into every sentence I have written.

This book is divided into three parts. In the introductory chapter, I give a brief overview of the history of building in Africa, the existing literature, and how this journey came about, to the best of my knowledge and understanding. The 'Diary' records 47 selected days in various locations that offer interesting facts about construction in Africa or were particularly eventful. The 'Considerations' – seven short texts at the end of the book – explore construction topics that caught my attention during the trip and continued to preoccupy me, prompting further research.

I hope that my writing efforts have resulted in something informative and worth reading. If so, it is because I was able to work with excellent support. First and foremost, I would like to thank the many civil engineers in Africa who took the time to meet with me and willingly provide me with information.

I am particularly pleased with the contributions from African colleagues, whose short essays on construction in their countries are a valuable addition to this book: Elsabe

Kearsley (South Africa), Philemon Arito (Namibia), Senthil Kumaran (Zambia), Katelula Kaswaga (Tanzania), Francois Nsengiyumva (Rwanda), Henry M. Alinaitwe (Uganda), Abebe Dinku (Ethiopia), Amged Abdelatif (Sudan), and Salah El-Metwally (Egypt).

This book came about because publisher Philipp Meuser agreed to include it in the DOM publishers programme. Björn Rosen is more than just the publishing director there; he was my coach and motivator. I would like to thank Uta Keil for her careful editing, which put the finishing touches on the text. Ursula Baus checked my manuscript beforehand; she was tough on me and pointed out the worst mistakes (and they weren't just commas) in good time and with good intentions. Photographer Wilfried Dechau was kind enough to edit my photos and get the best out of them. I entrusted the design of this book to my sister Sibylle and her firm, Moniteurs – with great success! Anna Gaißmaier masterfully brought the text and images to life in colour, and Annick Rietz is responsible for the fabulous data graphics. Moniteurs worked alongside Nicole Wolf, the graphic designer at DOM publishers.

I am very grateful to the Technische Universität Berlin (TU Berlin) for granting me a research semester. During my absence, my colleagues Karsten Geißler, Volker Schmid, and Alex Hückler ensured that teaching and research continued smoothly at the TU. Max Dombrowski standardised the bibliography and took care of the extensive data collection. Niika Dudzus and Haitham Al Abdulrazzak helped me with the list of bridges over rivers in Africa. Johannes Giebel from the geochemistry department and Wilfried Walkowiak and Ralf Lagerstein from the testing laboratory examined my rock samples from Rwanda, and Colette Howell assisted me with the initial travel planning.

My partners at schlaich bergermann partner also gave their immediate approval when I told them about my travel plans. The Berlin team, and especially Sven Plieninger and Michael Werwigk from Stuttgart, covered for me during my absence when there was an emergency. Fabian Gross reviewed the considerations regarding solar energy, and Monika Jocher provided helpful comments on my texts time and again. I had extremely stimulating and informative conversations with Ulrich van der Heyden about construction in the colonial era, with Simone Stürwald, Fabienne Hoelzel, and Luyanda Mpahlwa about African housing construction, and with Simon Züfle about China's influence in Africa.

Last but not least, I would like to thank Massimo, without whom I would never have been able to make this trip, and my wife Paloma Lasso de la Vega, who joined us twice and embarked on this adventure with me in Cape Town, Kenya, and Tanzania.

1 Introduction

Fig. 1 Abandoned rest stop on the A109 between Mombasa and Nairobi, Kenya

Introduction

Should we have turned back? Would it have been better to just keep driving and not get out to photograph this rare bridge? This is an 'engineering safari' that turned into a 'Cape to Cairo' adventure trip. It was no Robinson Crusoe story; we were too rarely alone and seemingly lost. And although things got dicey at times, we wouldn't want to forget a single day. Not the mud and constant rain on the way to the remote construction site of the continent's longest cable-stayed bridge on South Africa's Eastern Cape. Or the scorching hot sand in the empty Western Desert between Libya and Egypt's bustling capital of Cairo, which, with its 20 million inhabitants, is creeping ever closer to the pyramids.

Africa – Personal

Africa fascinated me even in my youth. For me, this continent has since been associated with a foreign yet attractive zest for life. The primary school in Stetten im Remstal is named after Karl Mauch (1837–1875) – "His greatest discovery was the ruins of Zimbabwe ..." [1]. I was a pupil there for four years, but at the time I didn't give much thought to the man who gave the school its name, who was born in my home village. More influential were the adventure books by Friedrich Wilhelm Mader (1866–1945), the "Swabian Karl May", which my grandmother in Stetten, who was a pastor's wife, had bought for me in antiquarian bookshops. As a child, I devoured *Kilimanjaro* [2] and *Die Flucht aus dem Sudan* [3] *('The Escape from Sudan')*. The works of this Protestant pastor are now referred to as colonial revisionist 'East African' novels. At the time, I did not even begin to realise that the adventures of the cunning followers of the Mahdi, who rebelled against the British in Khartoum, and the 'brave' German soldiers with their self-sacrificing askaris in what was then German East Africa, could be classified in this way.

In 1977, the enthusiasm I acquired through reading was enough to convince two school friends to join me on a trip to Sudan during the summer school holidays and to persuade my parents not to forbid their 17-year-old son from going. My father even gave me 100 Deutschmarks "for emergencies". We climbed the Pyramid of Cheops in Cairo, crossed the Aswan Reservoir on a boat despite being weakened by amoebic dysentery, and travelled by train across the Nubian Desert. Reaching Khartoum marked the end of the line for us. At the end of the 1970s, there was also unrest in Darfur and flooding in southern Sudan. We couldn't go any further and had to turn back.

In 1983, the lecture-free period was still called semester holidays, and it allowed me, a not-so-diligent student at ETH Zurich, to embark on a three-month backpacking trip to the Africa of Joseph Conrad and V.S.Naipaul. By hitchhiking and taking buses and trains, I travelled from Bangui in the Central African Republic – still suffering from the consequences of the reign of the self-crowned Emperor Bokassa I – to Zaire under the empire of Mobutu Sese Seko. Bizarrely, Mobutu named the country, the river, the currency, and even the beer 'Zaire' at that time. From there, I travelled via Rwanda to the island of Zanzibar. For a civil engineering student, the remains of Belgian colonial architecture in Zaire demonstrated the sad transience of structures in the jungle. Even then, I noticed that Chinese construction companies were entrusted with road construction in Rwanda.

Since then, I have travelled to Africa repeatedly for work. In 2006, for example, I went to South Africa at the invitation of the local steel construction association to give lectures at the universities in Stellenbosch and Pretoria. The contacts I made at that time were very helpful starting points for the 2022 trip.

Our consultancy, schlaich bergermann partner, is also occasionally involved in projects in Africa. For example, we were able to design the roof of the National Stadium in Abuja, Nigeria, for the 2003 All-Africa Games, and we were commissioned to design the roofs of the stadiums in Cape Town, Port Elizabeth, Durban, and Johannesburg for the 2010 FIFA World Cup in South Africa. [4] I was not directly involved in that project and had not yet found an opportunity to visit the South African stadiums. The 2022 trip, therefore, began with a visit to 'our' lightweight, long-span stadium roofs. In 2009, we were in close contact with Christoph Schlingensief and Francis Kéré regarding the planning of the festival theatre in the Opera Village Africa in Burkina Faso. However, after a few sketches and initial calculations, this project came to a standstill because its driving force, Schlingensief, passed away in 2010. Our consultancy has been designing solar power plants for over 40 years. One of them, the Kuraymat trough power plant, has been producing solar energy in Egypt since 2011.

Approach Through Books

The multi-volume *General History of Africa* by UNESCO, which occupies considerable shelf space, is considered the recognised standard work on the history of Africa, not least because both black and white historians contributed to it. [5] The fascinating and complex history of the continent in general is also summarised in an understandable and detailed manner in John Reader's *Africa, A Biography of the Continent* [6] and in Kevin Shillington's textbook *History of Africa* [7]. My bookshelf also includes the British work *Colour, Race and Empire* by A. G. Russell from 1944 [8] and *How Europe Underdeveloped Africa* by Walter Rodney from 1972 [9] – books that reflect the views of their respective times. A recent publication, *Deutschkoloniale Baukulturen (German Colonial Building Cultures)*, is a very readable volume on building in African countries during the colonial era. [10]

There is a wealth of travel literature available. In 1983, I set off with what was then the only book on Africa published by Lonely Planet: Geoff Crowther's *Africa on the Cheap* [11], which covered all the countries on the continent for backpackers in 400 pages. Today the Lonely Planet website offers around 20 travel guides for Africa with thousands of pages of coverage.

Anyone interested in learning about architecture in Africa will now find a reasonable selection of books on the subject. The magnificent seven-volume *Architectural Guide Sub-Saharan Africa* by DOM publishers [12], organised by country, guides travellers to the most important buildings in each location. The map of Africa in the first volume shows "Architecture Schools in Sub-Saharan Africa" and helped me find universities and thereby influenced my itinerary. Where architecture is taught, civil engineers are never far away. In 2022, the weighty and beautifully designed book *African Modernism* [13] was published, illustrating "the Architecture of Independence" in five countries on the continent. The 2023 Architecture Biennale in Venice, curated by Lesley Lokko, was also dedicated to Africa: "Firstly, Africa *is* the laboratory of the future." [14]

However, if you look for publications on engineering structures in Africa, on bridges, building structures, and road construction, you will be disappointed. There is no book that even asks about "the art of structural engineering" in Africa, as David Billington did for Europe and the US in his classic *The Tower and the Bridge* [15]. Apart from civil engineering structures in Egypt and South Africa, little has been published. Despite the major construction tasks ahead, there is hardly any literature on civil engineering in Africa. There is a need to catch up.

The Civil Engineer Can: Numbers

Construction in Africa fascinates me. Africa is so vast – and so unknown in Europe. This is especially true when it comes to civil engineering. Below is a brief overview, a comparative snapshot from the perspective of a civil engineer: but that is all it can be.

As an engineer, I initially approached Africa through numbers: With an area of around 30 million square kilometres, Africa is seven times larger than the European Union. This is clearly visible by glancing at the globe, yet flat world maps with Mercator projection make Europe appear much too large. In 2020, around 1.3 billion people lived in Africa, which corresponds to a population density of around 43 inhabitants per square kilometre Fig. 1. With a projected population growth to around 2.5 billion by 2050, this will rise to 83 inhabitants per square kilometre Fig. 2. Even if we exclude the Sahara, which covers almost 10 million square kilometres and makes up around a third of the continent, the figure would only be 125 inhabitants per km^2. This is only slightly more than the population density of the 450 million EU citizens, which is around 110 inhabitants per km^2, and only 60 per cent of Germany's population density. So, there is space, but an incredible amount of construction is needed.

This is because the approximately 1.2 billion people who will be born in Africa over the next 30 years represent almost three times the population of the EU. The needs of these people, such as food, health care, and education, must be met. Not least, this involves the development of infrastructure: housing, schools, and hospitals must be built, and roads, railways, and bridges must be constructed. Imagine if, in 30 years' time, the entire European building stock were to be rebuilt three times over in Africa! Approaching this task as we have done up to now will probably fail due to a lack of material resources. The volume of concrete and steel – let alone wood – will not be available. So, I am not thinking about building car-friendly cities. And it will probably not be enough to simply switch to rail transport. What is needed is a completely new type of construction and transport system that we cannot even imagine yet.

Mission impossible? In any case, we civil engineers cannot avoid dealing with what is probably the world's biggest construction task at the moment, an incredible challenge. We have to get started because the pressure to build and the housing shortage are already enormous – during my trip, I heard one thing everywhere: "We need decent and affordable housing" (→ Considerations – Residential Construction).

What are the prerequisites? And what is the current state of affairs? In her book *The Challenge for Africa* [16], Nobel Peace Prize winner and founder of the NGO The Green Belt Movement, Wangari Maathai from Kenya offers a compellingly clear and objective overview. She is dissatisfied with the status quo and calls for "... a revolution in leadership – not only from the politicians who govern, but from active citizens who place their country above the narrow needs of their own ethnic group or community". A functioning state is crucial. There is no shortage of solutions, proposals worth considering, and far-sighted ideas from outside. In *Africa First!* [17], Martin Schoeller and Daniel Schönwitz see the solution in the introduction of a primarily social market economy on the continent, in a "European economic model" that contrasts with "US digital capitalism" and "Chinese-style state capitalism". In his book *Exodus* [18], economist Paul Collier describes a "social model", a combination of the "four pillars for prosperity" – namely institutions, rules, norms, and organisations – to enable prosperity to develop. Not all African countries have these pillars in place. A wide variety of causes are cited to explain the problems that many African countries still have to solve: the aftermath and wounds of the inhuman slave trade, the yoke of colonialism, the deep scars of apartheid in South Africa, the influence of superpowers through their proxies, misguided development aid, debt traps set by unfair lenders, ethnic structures incompatible with democracy, and incompetent and corrupt politicians. All these components play different roles, with varying degrees of importance depending on the country. I encountered corruption frequently during my trip, and it struck me that wealth is even more unevenly distributed than it is in my part of the world.

Yet Africa is so rich in people, farmland, and natural resources. In Botswana, Namibia, and South Africa, for example, the De Beers consortium – founded by Cecil Rhodes in 1888 – mines diamonds. The world's largest diamond producer has for years limited itself to selling "conflict-free diamonds", but 4 per cent of "blood diamonds" are still on the market [19]. Botswana holds a 15 per cent stake in De Beers, and diamonds account for a major percentage of the country's exports. In Zambia, large copper deposits were discovered in the so-called Copperbelt shortly after independence, and exports of this metal make up around 75 per cent of the country's raw material exports. Other countries have more diversified economies. In Namibia, for example, there are uranium mines; gold, copper, and zinc are also mined, and offshore oil and gas have been discovered. Above all, however, Namibia is rich in sun, wind, and deserts. I am convinced that Africa's abundance of sunshine can cover its own energy needs and that good money can be made by supplying the surplus to Europe. (→ Considerations – Solar Energy).

A Brief History of Construction in Africa

The continent was largely under African rule until the early 1870s. European and Arab trading posts were limited to the coasts and river courses. Then began the European 'Scramble for Africa', Europe's fatal colonial race for Africa. Today's national borders largely correspond to those of the colonies of the early twentieth century Figs. 3–4. They are a consequence of the so-called Congo Conference convened by Otto von Bismarck in Berlin in 1884/1885. "Its final document, the Congo Act, formed the basis for the division of Africa into colonies in the ensuing race for Africa." [20] Ethiopia is considered the only country that has always been independent, except for a brief occupation by Italy from 1935 to 1941. Following the Second World War and mostly after 1960, almost all African countries gained independence, but this did not necessarily bring freedom, stability, and prosperity everywhere. The situation is particularly sad in the green lung of Africa: the Congo. Since gaining independence from Belgium in 1960 and the assassination of its first Prime Minister, Patrice Lumumba, in 1961, the country now known as the Democratic Republic of Congo has scarcely found peace.

It is obvious that the country borders are not based on the ethnic groups of the respective region. A particularly curious example is the Caprivi Strip, a thin strip of land that juts out from north-eastern Namibia between Angola and Zambia in the north and Botswana in the south. The German Empire bought this piece of land from England in order to build a German railway line from German South West Africa, as Namibia was called during colonial times, to German East Africa, now Tanzania. Every country in Africa is home to a multitude of ethnic groups. A few names may be familiar to Germans in particular because we are finally admitting that the slaughter of the Herero and Nama in Namibia by German colonial troops was genocide. Across the rest of the world it has been impossible to ignore the news reports that the Tigray, Oromo, and Amhara peoples in Ethiopia have been engaged in bloody warfare in recent years. To come to terms with these stories and understand them, I would need more time than a passing visit allows. In Zambia alone, 72 ethnic groups [21] are recognised. Referring to 'tribes' rather than 'ethnic groups' is often considered politically incorrect in Africa, but not everywhere. When we checked into a hotel in Tanzania, we had to state our 'tribe'. The guestbook of the Medy One Lodge in Nyakanazi, where we stayed in June 2022, now includes an entry from a Swabian and a Mantovan. And it's true: I feel I belong to the 'tribe' of the Swabians.

Anyone involved in construction in Africa will find no shortage of evidence of 'pre-colonial' building. Of course, sophisticated construction was already taking place in Africa before the Greeks and Romans reached the north of the continent, before the Arabs came to the east coast, and before the first Europeans landed on the coasts of southern Africa at the end of the fifteenth century. The regions along the northern Nile, from Sudan to Egypt, which had been made fertile by flood mud and were connected to the other empires of the Middle East, gave rise to advanced civilisations as early as the fifth millennium BCE. Anyone travelling the approximately 3,000 kilometres along the Nile from Cairo to Khartoum can see for themselves this exchange and intermingling of cultures, which continues to this day. The multitude of buildings constructed by the Egyptians, Nubians, Greeks, Romans, Ottomans, French, and British is both overwhelming and confusing. The incomparable buildings from the time of the pharaohs are not limited to the temples and tombs between Giza and Abu Simbel. South of the Aswan Dam, you can also find Egyptian temples and beautiful Nubian pyramids, such as those at Meroe near Khartoum. Unfortunately, in the 1830s, the Italian 'archaeologist' Giuseppe Ferlini blew off most of their tops in search of tomb treasures.

In predominantly Christian Ethiopia, the churches of King Lalibela, carved out of rock around 1200, are impressive. According to legend, they were intended to replicate Jerusalem in order to spare pilgrims the arduous journey to the Holy Land. These monolithic structures with vaults, supports, beams, and lintels are magnificent examples of subtractive building. The load-bearing behaviour of these stone structures in the Amhara region has not yet been studied in detail. Long before Christ, the city of Aksum, now a World Heritage Site located further north in Tigray, was an important trading hub for goods between Asia and Africa. The imperial city of Gondar was Ethiopia's first permanent capital from 1636 to 1885. In the palace district there, you can marvel at Fasilides Palace, whose 'African eclectic' architecture combines Indian, Moorish, and Portuguese elements. Much less well known are the completely unrestored seventeenth-century stone arch bridges located just outside the city gates. These Ethiopian treasures are probably the oldest bridges south of the Sahara.

Unfortunately, the largest pre-colonial stone structure south of the Sahara – Great Zimbabwe – was not on our route because we had planned university visits on a different route. This ruined city in Zimbabwe, with its impressively high mortarless walls, experienced its heyday between the eleventh and fifteenth centuries, owing its wealth to the gold trade with Arabia and Persia. Karl Mauch, the German explorer and geologist who also worked as a gold prospector, deserves credit for its discovery. He was the first to examine the ruins intensively in 1871 and, after months on site, returned home suffering from malaria and made the city known to the public in

Fig. 1 Population density in the 12 countries visited

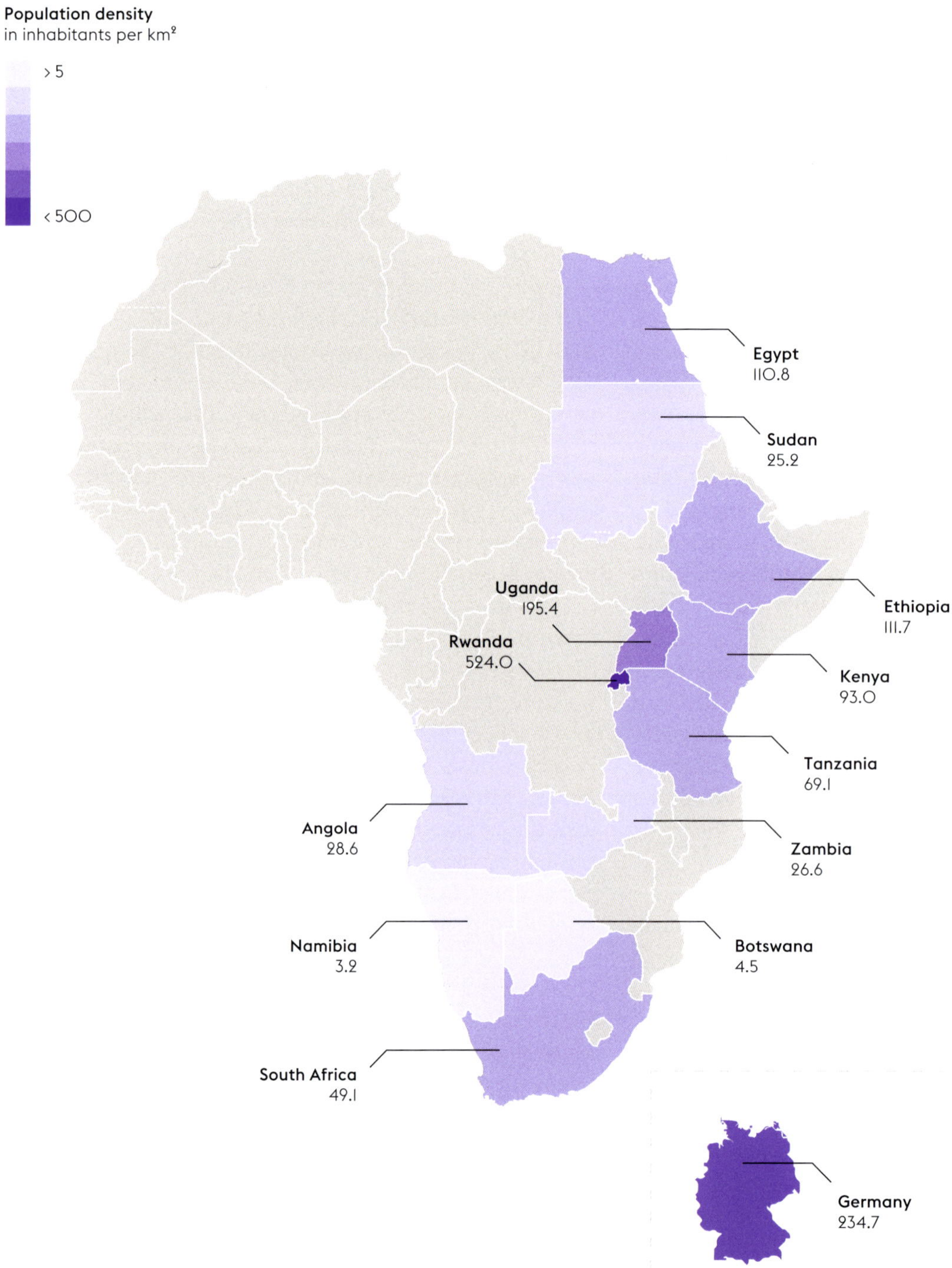

Source: World Bank | Wikipedia, see data sheet from page 220 onwards

Fig. 2 Population growth and GDP per capita in the 12 countries visited

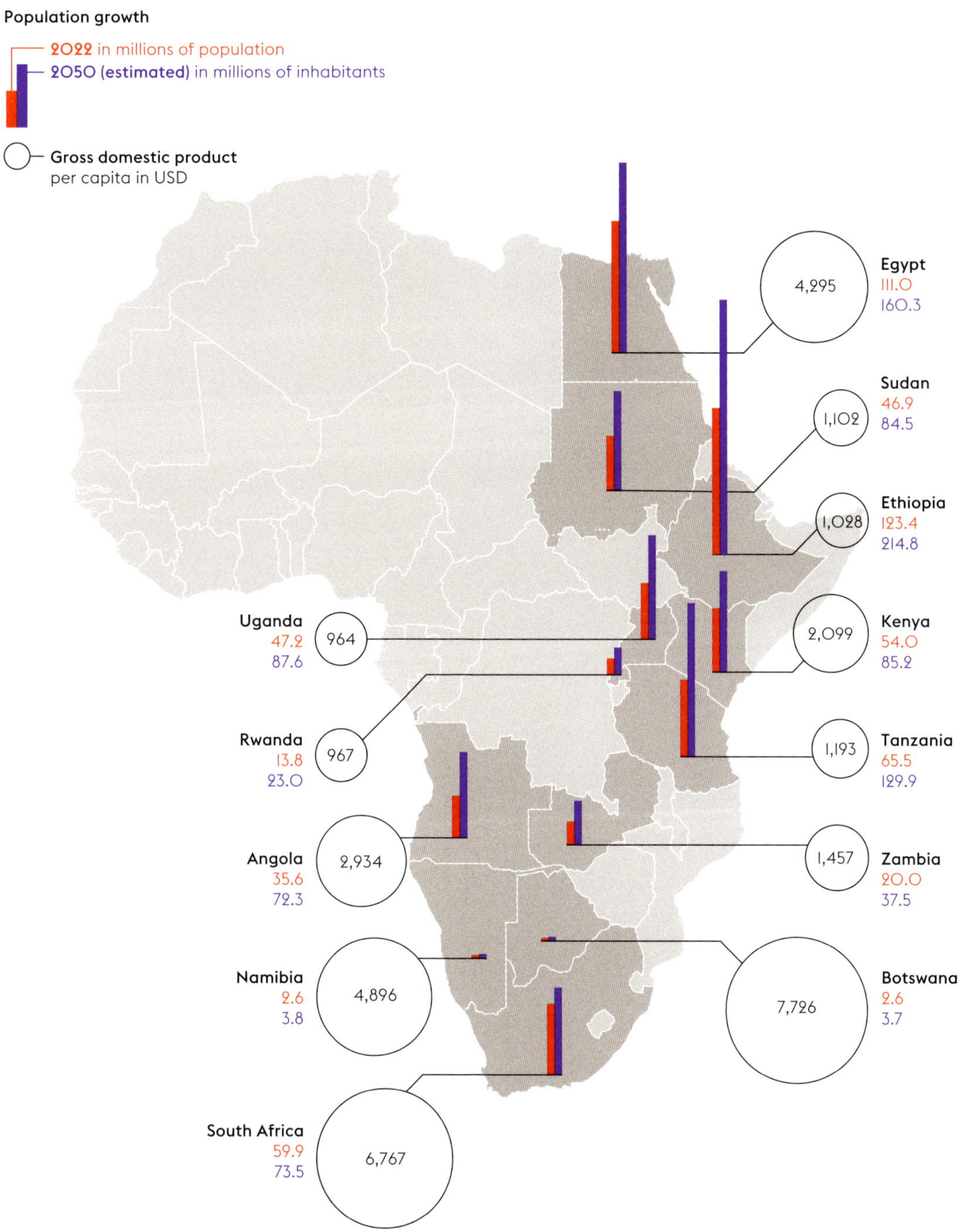

Source: World Bank, see data sheet from page 220 onwards

Fig. 3 Africa: National borders as of 1914

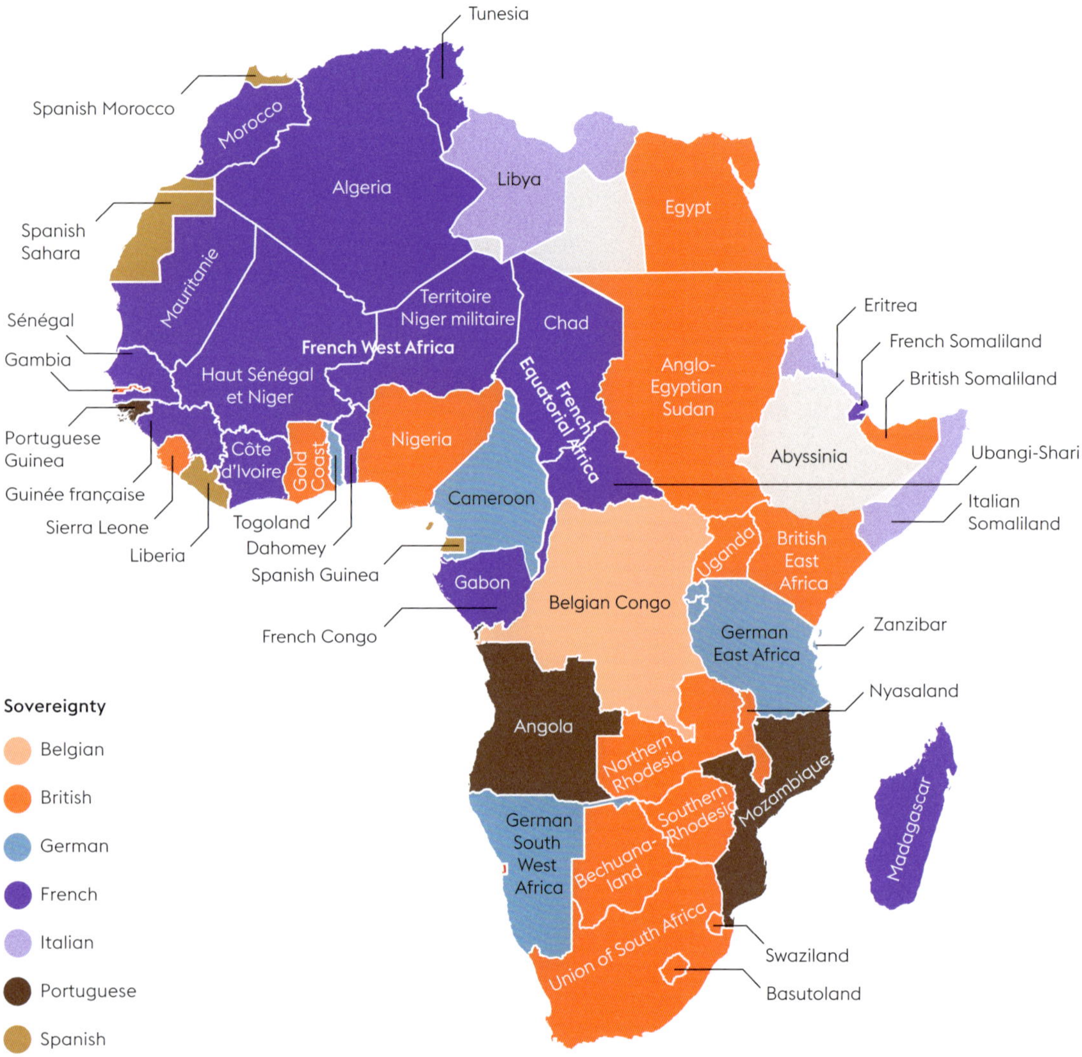

Source: Representation based on Hogweard, Africa map 1914, wikimedia commons, 2023

Fig. 4 Africa: National borders as of 2022

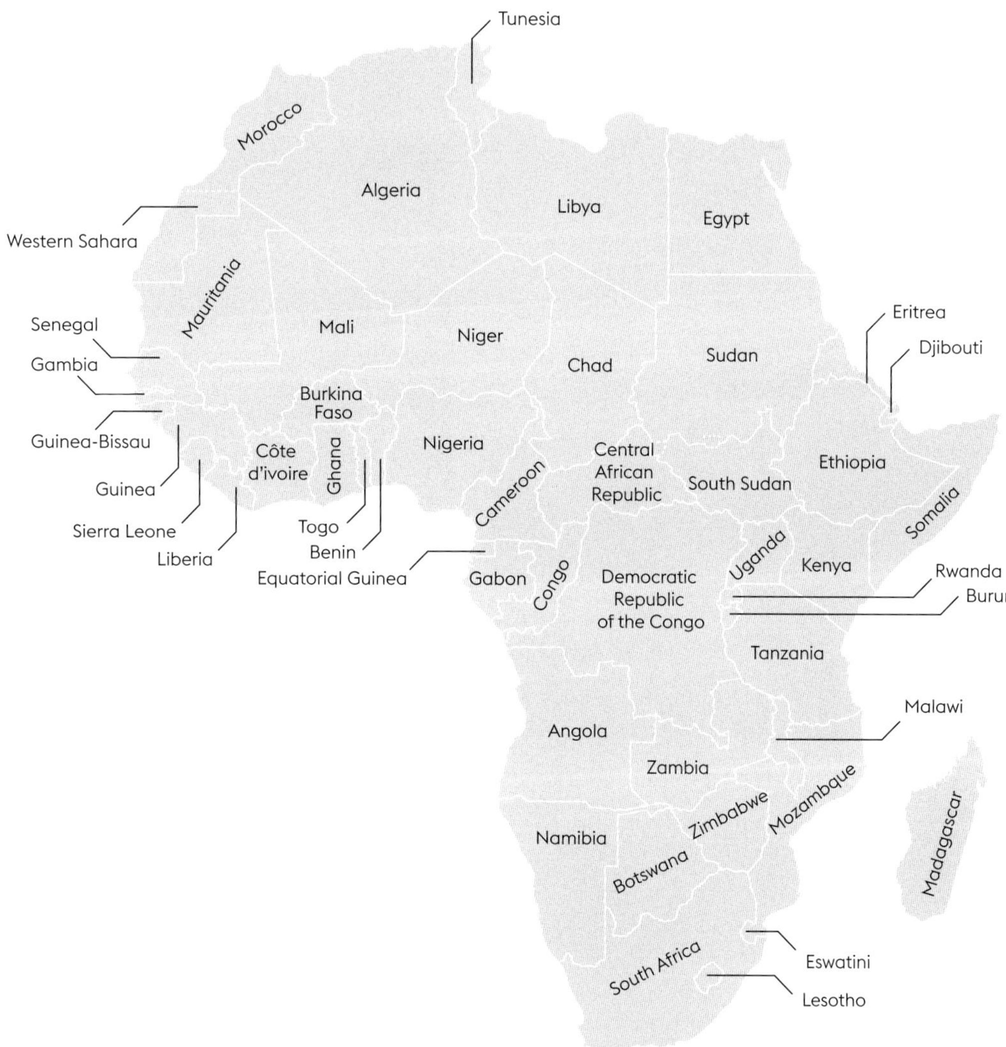

Source: Own representation based on Google Maps, April 2024

5

6

7

8

9

10

11

12

13

14

15

16

Figs. 5–16 Indigenous self-build in Namibia (5), Botswana (6), Angola (7–8), Zambia (9), Tanzania (10), Kenya (11–12), Ethiopia (13–14), Sudan (15), and Egypt (16)

17

18

19

20

21

22

23

24

25

26

27

28

Figs. 17–28 Colonial buildings in South Africa (17), Namibia (18–19), Angola (20), Zambia (21), Tanzania (22–23), Uganda (24), Kenya (25), Ethiopia (26), Sudan (27), and Egypt (28)

Germany. [22] Great Zimbabwe is not an isolated case. The remains of more than 500 stone palaces have been documented between South Africa, Zimbabwe, and Botswana. [23]

West Africa also boasts impressive pre-colonial buildings. Just think of the multi-storey mud buildings of the Dogon in Mali. Most countries in the Sahel region cannot currently be visited because Islamist terrorist groups such as ISIS and Boko Haram as well as marauding gangs are wreaking havoc there.

Outside the early centres of construction highlighted here, however, large parts of sub-Saharan Africa were simply too sparsely populated and too poor to enable the emergence of a 'great' engineering culture. Hunters and gatherers, as well as cattle breeders and farmers, do not build temples and palaces. Their houses, vernacular or indigenous buildings with walls made of clay, wood, and stone constructed locally and mostly with thatched roofs, can be found everywhere from South Africa to Sudan. Adapted to the local climatic and geological conditions in each region, these houses are, at least at first glance, prime examples of sustainable construction. They embody climate-friendly building with local natural materials that are returned to nature at the end of their life cycle without producing any waste (→ Considerations – Concrete) Figs. 5–16.

Despite all of this, we know comparatively little about pre-colonial construction in Africa. This is because many African cultures had no written language, which is why research relies on incomplete oral traditions and archaeological finds. The first written documentation was only created during the colonial period. Unfortunately, this means that Africans who want to study their history today have to travel to archives in Europe. This also applies to the history of slavery, which began as trans-Saharan trade before spreading across the Indian Ocean, reaching its horrific peak in the trans-Atlantic slave trade. In his book *Geschichte der Sklaverei ('History of Slavery')*, Andreas Eckert describes "the Atlantic slave trade as a radical break in the history of Africa". [24] Of course, it also had an impact on construction in Africa.

African societies, armed only with traditional weapons and guided by their beliefs, were poorly equipped to oppose European settlers with firearms and artillery as well as missionaries motivated by religious conversion. Some missionaries may have meant well, but they caused a "deep cultural inferiority complex" [16] that continues to have a damaging effect to this day. Only a few African peoples, such as the Ethiopians, managed to keep the Europeans at bay, or, like the Maasai in Kenya and Tanzania, were able to preserve their way of life throughout the colonial period.

The colonial settlers took the customs and the types of construction of their respective homelands with them to foreign lands. Just as patterns of consumption persist to this day – southern Africa relying largely on South African wine, Angola on Vinho Verde, and Rwanda on Bordeaux, despite the use of English now – so too do transportation norms, such as driving on the side of the road determined by former colonial powers: left in South Africa, right in Angola, left in Zambia, and right in Rwanda. The university in Khartoum, originally the Gordon Memorial College (1902), stands out for its eclectic Gothic and Romanesque style; similarly, Swakopmund, Namibia, feels like a remnant of nineteenth-century Germany. If it were not for the palm tree growing in front of the Lutheran church there, the setting could easily be mistaken for the southern German state of Baden-Württemberg. The railway station in Kigoma, built in 1915, could also just as easily be located in Germany. It is the terminus of the Tanganyika Railway, the "greatest technical legacy of the German colonial era in Tanzania" [25], which at that time connected the port of Dar es Salaam with Lake Tanganyika Figs. 17–28.

During our stay in Kigoma, the Liemba was anchored and once again awaiting repairs Fig. 23. I saw this beautiful oldtimer here before in September 1983, and I hope that this, one of the oldest still-operating passenger ships in the world, so important to the lake's residents, will soon be able to set sail again. It is the only remaining former German warship from the First World War. [26] After being sunk by German troops during their retreat in 1916 and raised by the British in 1924, it served as a passenger ship on Lake Tanganyika until 2018.

Fig. 29 The Fiat Tagliero petrol station in Asmara, Eritrea

29

Just as the Germans transported this ship to Africa in prefabricated parts and assembled it there, the British brought prefabricated steel bridges to the continent. Steel truss bridges like the White Nile Bridge (Old Omdurman Bridge) in Khartoum or the historic railway bridge over the B6 from Windhoek, Namibia, to the international airport, are a common sight for travellers in the former British colonies across the continent.

Today, the Italian art deco architecture in Asmara, the capital of Eritrea, seems almost surreal. [27] Mussolini wanted to create a 'Seconda Roma' here. At every turn, walkers encounter well-preserved but far too large and therefore empty grandiose cinema buildings such as the Cinema Roma, Cinema Impero, Cinema Odeon, and Cinema Capitol. The most spectacular, albeit structurally unremarkable, is undoubtedly the wing-shaped roof of the Fiat Tagliero petrol station, which became a UNESCO World Heritage Site in 2017 Fig. 29. In contrast, the construction of the 200-kilometre-long railway line from the port city of Massawa – the 'Pearl of the Red Sea' – to Asmara (2,300 metres above sea level), was a technical masterpiece when it was completed in 1911. On the last steep section before Asmara, 24 tunnels and 65 bridges were built on a stretch of only 48 kilometres. [28] When I visited the country in 2012/2013, however, this section was in a deplorable state, only partially passable and then only for tourists. Reactivating this line, which carried up

30

Fig. 30 Modernism in Nairobi, Kenya

to 38 trains a day during the colonial era, would greatly benefit this poverty-stricken country, which has been ravaged by its founding fathers, the Eritrean People's Liberation Front.

For most African countries, the colonial era ended two generations ago, but the post-colonial period has been difficult despite the initial optimism. From 1960 to 1990, economic growth in many African countries was very low or even negative, with a few exceptions such as Botswana (annual growth of 6.5 per cent [16]). [29] This was also due to the fact that the countries of the Organisation of African Unity (OAU; the African Union since 2002), founded in 1963, were drawn into the Cold War blocs and became involved in proxy wars. This had an impact on construction. Because Zanzibar officially recognised the 'German Democratic Republic' (GDR) as a state in 1964, Walter Ulbricht, who also built the Berlin Wall, expressed his gratitude with "prefabricated buildings, which were intended to bring the joys of socialism to the island's inhabitants as 'modern living space'" [30]. Since the collapse of the Soviet Union and the end of the Cold War, at least these influences, which weakened the idea of a united Africa, seemed to have disappeared into oblivion. However, the war in Ukraine shows that the old ties still exist. Many African countries, especially South Africa, merely abstained from voting on UN resolutions against Russian aggression in Ukraine because of their dependence on Russia or their aversion to the US. Only a few countries in southern Africa, such as Kenya, took a stand against Russia's war of aggression by voting in favour of these resolutions.

"If the colonial powers had promised education but not delivered, if they had promised minimum standards of living that had not come to pass, now was the time for the new leaders. They pledged to fulfil these hopes and expectations. Hence modernisation became the leitmotif of most, if not all, decolonised countries and their presidents," writes architect Manuel Herz. [13] The euphoria of the early years after independence is clearly reflected in the architecture of capital cities such as Nairobi. Today, the Central Business District (CBD) of Nairobi looks like an open-air museum of 1960s and 1970s architecture Fig. 30. Among the masterpieces of modernism and the beautiful exposed concrete buildings in the centre, two structures stand out in the truest sense of the word: the 105-metre-high Kenyatta International Conference Centre by Karl Henrik Nøstvik and the free-standing tower of the Cathedral Basilica of the Holy Family by Dorothy Hughes. Of greater interest to civil engineers is the Wakulima Market, built in 1966 by the Chief Architect of Nairobi Town Council. It features 60 tall, free-standing concrete mushroom columns, each composed of four strangely bent hyperbolic paraboloids. The influence of Félix Candela's resource-saving concrete shell structures thereby extended not only to Germany, where it was successfully emulated by GDR shell expert Ulrich Müther, but also to East Africa. Not much has changed in the CBD over the past 40 years. Today, construction in Nairobi takes place in less congested parts of the city.

Comparable exposed concrete masterpieces can also be found in Lusaka and Kampala because Zambia (1962) and Uganda (1964) gained independence at almost the same time as Kenya (1963). Few of the young states had universities or locally trained architects and engineers at the time of independence. Architects from the recently departed colonial powers were certainly not always at the top of the new states' wish lists. It is therefore not surprising that architects from Scandinavia and Israel often shaped the architecture of the early days of independence.

Post-war modernity, modernism, the International Style, and brutalism have also left their mark on Sudan. Major cities there are characterised by buildings such as the parliament building in Omdurman, a huge exposed concrete structure dating from 1972. Looking at contemporary buildings in Khartoum, such as the Corinthia Hotel, known as 'Gaddafi's Egg', one has to agree with Sudanese architect Omer Siddig Osman: "Architectural production in the country is today characterised by imitations and rootlessness. Attempts at emulating a regional approach are usually misguided or politically motivated. The economic aspirations that resulted from the oil wealth in 2002 initiated a trend for buildings modelled on those in the United Arab Emirates and Malaysia." [31] This sense of regret expressed in Sudan can be applied to the whole of Africa. Remarkable exceptions, such as the buildings by African architects Mphethi Morojele,

Diébédo Francis Kéré, David Adjaye, Olajumoke Adenowo, and Kunlé Adeyemi, are highlighted by Philipp Meuser and Adil Dalbai in their book *Theorising African Architecture: Typologies, Buildings and Urbanism.* [32] Throughout Africa, women are playing an increasingly important role in architecture: 50 influential African female architects are featured on the *Africans Column* internet portal, which is dedicated to art, architecture, and design in Africa. [33]

A quick search on the internet is usually enough to identify the architects of important African buildings from independence to the present day. However, as is so often the case here, there is usually no information about the engineers. Only a few new road and bridge constructions undertaken during the independence era are known. Instead, colonial infrastructure buildings fell into disrepair, certainly in part due to the shortage of African engineers to manage the corresponding planning and maintenance tasks. Local self-build and traditional architecture can perhaps be transferred to modern building construction, but not to today's large-scale bridge construction. That is why bridge design and construction in Africa, at least for larger bridges, is often still imported from Europe, Japan, South Korea, and increasingly China (→ Considerations – Bridge Construction).

China has been another player in Africa for several decades. Today its influence can no longer be overlooked. At every turn, travellers encounter ports, roads, bridges, railway lines, and airports built by the Chinese (→ Considerations – China).

Travelling by Car

My plan was to learn about civil engineering between Cape Town and Cairo. However, it was not until our second attempt that the plan was successful. When we wanted to travel in March 2020 – after our car had already been shipped to Africa – airports, borders, and universities were closed due to the Covid-19 pandemic, and all our travel plans were thrown into disarray. It was not until 2022 that we were able to set off with new visas, vaccinations, and permits. We spent 150 days travelling in our Land Cruiser, burning 6,000 litres of petrol – admittedly not exactly sustainable – and covering 30,000 kilometres, during which we were able to experience *('erfahren')* Africa on the continent's roads in the truest sense of the word.

When we began our journey in April 2022, southern Africa was rather rainy and cool. It also rained a lot in northern Ethiopia when we were there in July and there was no shortage of water there. In the Nubian Desert, on the other hand, the thermometer showed 50° Celsius at midday in August, and southern Ethiopia, Somalia, and northern Kenya were experiencing extreme drought. The last rainy seasons had failed. Tsavo East National Park in Kenya was bone dry, and the animals gathered at the dirty green remnants of the Tsavo River. As early as October 2000, Peter Scholl-Latour wrote: "Kenya is suffering from the worst drought in half a century." [34] Nothing new, then, but according to the UN Drought Report 2022, the number and duration of droughts have increased by 39 per cent compared to the previous two decades. Climate change is hitting Africa hard, and once again the United Nations has to distribute food and water.

What image of Africa is conveyed in Germany? Anyone who reads Peter Scholl-Latour's reports in the books *Afrikanische Totenklage ('African Dirge')* and *Mord am großen Fluß ('Murder on the Great River')* [35] comes away with the impression of a continent permanently plagued by murder and manslaughter. Just like his fellow journalist Ryszard Kapuscinski, Scholl-Latour was always present when the bloodiest upheavals took place at the end of the colonial era and in the decades that followed. We have adopted these images, and indeed, even today, religious fanatics in Nigeria, power-hungry generals in Sudan, marauding gangs in South Sudan, and cattle breeders and farmers in the Sahel region plagued by overpopulation and water shortages are triggering bloody conflicts on the continent. Countries such as Congo, Burundi, South Sudan, and Nigeria, as well as most of the Sahel region, are currently too dangerous for private travel.

On the ground, Africa seemed much more stable, peaceful, and safe than I had imagined before my trip. Namibia and Botswana are safe countries to travel to and seem to have corruption under control. Rwanda is as clean as Switzerland, and Kenya held fair elections again in 2022. For the 'overlander' in Africa, a car accident is the greatest danger (→ Considerations – Roads and Railways).

Colleagues

My journey through 12 countries in southern and eastern Africa in 2022 revealed the incredible diversity of construction there. Perhaps the impressions and observations of a civil engineer can shed light on at least some aspects of construction and contribute to an understanding of civil engineering there.

My research semester on the topic of construction in Africa started out as a rather vague idea: to travel through Africa and see how construction is done there. It was clear that I wanted to meet as many construction professionals as possible. The trip was to take me to many engineering structures, such as bridges and towers, and architectural highlights. The list of universities I wanted to visit to learn about teaching and research there changed repeatedly during the trip. For example, on Friday 8 July, it was announced in Kenya that the following Monday would now be an Islamic holiday. Unfortunately, this meant that my planned visit to the University of Nairobi on that day had to be cancelled. On the other hand, a recommendation from a colleague in Kigali, Rwanda, enabled me to make an unexpected visit to INES Ruhengeri in the north of the country. At each university, I tried to talk to colleagues, get in touch with students, visit the laboratories, and conduct a standard interview, which was supplemented by a questionnaire sent to all interviewees in 2023. I was also able to visit a number of engineering firms, meet civil engineers, and get an idea of the projects they were working on and the situation in the local construction industry. In total, I gave around 20 lectures on bridge construction or the design and construction of supporting structures (→ Considerations – Universities).

I always asked how *Baukultur* was faring under the given circumstances, in the respective historical and cultural context. Are there signs of an African civil engineering approach that can meet the challenges of the above-mentioned 'mission impossible'? How are sustainability issues dealt with? Is there talk of 'the art of civil engineering' in the country? Are there any structures typical of the country? The answer to the last question was often: "No, we have the same standards!" However, pointing out how climatic conditions, wind loads, seismic accelerations, local materials and construction methods, as well as financial and political constraints affect the construction of bridges, roads, and houses, usually led to interesting discussions.

The idea for this book was born from the multitude of observations and impressions gained on this trip. However, it was only after intensive subsequent research, further discussions with colleagues from Africa and Germany, a dozen bachelor's theses at my own chair on the construction industry and bridge building in individual African countries, and a series of lectures at the Technische Universität Berlin on bridge building in Africa that my knowledge was supplemented as it is now presented here. In the summer of 2023, I asked several African university colleagues to write essays on civil engineering in their own countries. These short texts provide additional valuable insights and offer diverse perspectives on construction in Africa.

2

Diary

Fig. 1 Old Blue Nile Bridge (1908) in Khartoum, Sudan

End of the Journey
Start of the Journey
Tunis
Algeris
Rabat
MOROCCO
TUNESIA
Tripoli
ALGERIA
LIBYA
EGYPT
Cairo
WESTERN SAHARA
MAURITANIA
Nouakchott
MALI
NIGER
CHAD
SUDAN
Khartoum
ERITREA
Asmara
DJIBOUTI
Djibouti
Dakar
SENEGAL
GAMBIA
Banjul
GUINEA BISSAU
Bissau
GUINEA
Conakry
SIERRA LEONE
Freetown
LIBERIA
Monrovia
Bamako
Ouagadougou
BURKINA FASO
Niamey
BENIN
NIGERIA
Ndjamena
Abuja
TOGO
GHANA
CÔTE D'IVOIRE
Yamoussoukro
Accra
Lome
Cotonou
Addis Ababa
ETHIOPIA
CENTRAL AFRICAN REPUBLIC
SOUTH SUDAN
Juba
CAMEROON
Bangui
Malabo
Yaounde
UGANDA
SOMALIA
Mogadishu
EQUA TORIAL GUINEA
Kampala
KENYA
Libreville
GABON
REPUBLIC OF THE CONGO
RWANDA
Kigali
Nairobi
DEMOCRATIC REPUBLIC OF THE CONGO
BURUNDI
Gitega
Brazzaville
Kinshasa
TANZANIA
Dodoma
Luanda
ANGOLA
ZAMBIA
Lusaka
MALAWI
Lilongwe
Harare
ZIMBABWE
MOZAMBIQUE
MADAGASCAR
Antananarivo
NAMIBIA
Windhoek
BOTSWANA
Gaborone
Mbabane
Maputo
ESWATINI
Maseru
LESOTHO
SOUTH AFRICA
Cape Town

Diary

1.	South Africa	15 March–14 April 2022
2.	Botswana	14–28 April 2022
3.	Namibia	28 April–12 May and 17–19 May 2022
4.	Angola	12–17 May 2022
5.	Zambia	19–31 May 2022
6.	Tanzania	31 May–6 June and 24 June–1 July 2022
7.	Rwanda	6–10 June 2022
8.	Uganda	10–13 June 2022
9.	Kenya	13–24 June and 1–15 July 2022
10.	Ethiopia	15–31 July 2022
11.	Sudan	31 July–15 August 2022
12.	Egypt	15 August–6 September 2022

ZIMBABWE
NAMIBIA
BOTSWANA
Maun
Sowa
Orapa
Francistown
Omaruru
Okahandja
Gobabis
Windhoek
Phikwe
Serowe
Palapye
Shoshong
Mahalpye
Musina
Mavhuwa
Dzata
Mankgaile
Molepolole
Mochudi
Jwaneng
Gaborone
Ramotswa
Lobatse
Sun City
Klipgat
Phokeng
Pretoria
Fourways
Katlehong
Johannesburg
Ithuba
ESWATINI
Keetmanshoop
Lüderitz
Dithakong
Heilbron
Nkandla
Kimberley
Bloemfontein
LESOTHO
Pietermaritzburg
Durban
SOUTH AFRICA
Day 28, Msikaba Bridge
Day 25, Mthatha
Graaff-Reinet
Makhanda
Oudtshoorn
Cape Town
Genadendal
Alexandria
Day 1+14, Cape Town
Day 7, Stellenbosch
Hotagterklip
0 km
250 km

South Africa
15 March–14 April 2022

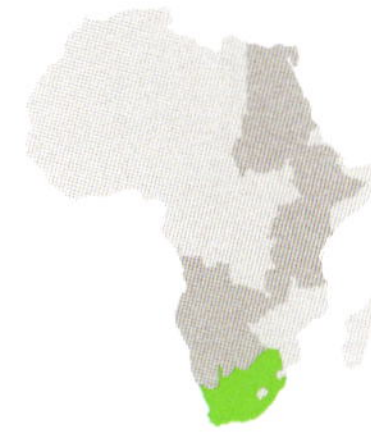

Who hasn't heard of the Cape of Good Hope, Table Mountain, Kruger National Park, or the Krugerrand? South Africa is one of the countries with the highest standard of living in Africa. It carries weight on the global political stage and can be described as a tourist magnet. The terrible years of apartheid were largely overcome without violence or bloodshed, and yet the country, with its 60 million inhabitants, does not seem to be doing well. Even the remaining five million white people of European descent are not all happy with the situation here. One million of them, mainly young people, have left the country in recent years. [1] They go to Europe or the US. This is because here in South Africa, anyone who can afford to lives behind high bulletproof walls, bars, and barbed wire. Security services patrol the streets, and in the evenings, the safest way to get around in public spaces is by car. South Africa is the most dangerous country on the trip. More on this later.

Day 1, 15 March – Cape Town

Arriving at Cape Town International Airport is a stress-free experience. The wayfinding is exemplary, and your luggage is typically already waiting for you on the carousel. You can pick up your hire car at the local Sixt branch and set off for Cape Town in a clean, new car. We drive along the four-lane, well-signposted Highway No. 2 to the chic villa district of Newlands. I hardly notice that we are driving under the same concrete precast bridge decks that are unfortunately common in our country, too. I have to concentrate on driving on the left, but by the time we arrive, I have almost got used to it.

We booked our accommodation online: the van Zyl family's newly renovated two-room garden house. White-painted walls and beams under the roof, simple wooden furniture with linen covers, white roses in a vase, and a direct view of Table Mountain under a blue sky from the sofa: better than the photos on LekkeSlaap! That's what Airbnb is called here. Our hosts are my age, and I chat in German with Debby, who is athletic, slim, and always well dressed. She is a piano teacher and went to a German school in Namibia. The jovial, already greying Neil is a solicitor, and he certifies copies of all my documents, and I remain grateful for this whenever I have to deal with official matters during my trip that require them. The couple lives in the main house, which has a large garden and pool, and readily includes us in their family life. They invite us to dinner and take us on a hike up Table Mountain that lasts several hours and is challenging for office workers like us. Cape Town offers tourists everything from dramatic landscapes at the Cape and kitesurfing on beautiful sandy beaches to the deindustrialised harbour area with seafood restaurants and the MOCAA, the Zeitz Museum of Contemporary Art Africa, which is housed in concrete silos spectacularly cut open by British architect Thomas Heatherwick Fig. 1.

The Bright Sky, the freighter travelling from Hamburg to Cape Town and carrying the container with our car – or rather expedition vehicle – is delayed and will not arrive

1

2

Figs. 1–2 The sawn-open concrete silos at MOCAA in Cape Town and the Bright Sky that will deliver our car

until 3 April Fig. 2. This gives me time to visit colleagues at the universities of Stellenbosch, Cape Town, and Pretoria and even make a detour by plane to the Namibia University of Science and Technology (NUST) in Windhoek, Namibia, where we will not arrive by car until 28 April.

Day 7, 21 March – Stellenbosch

Studying in a wine-growing region is possible at the renowned Stellenbosch University, 50 kilometres from Cape Town, where the focus is on engineering sciences. Civil engineers may regret that this university, which was reserved for whites during the apartheid era, does not have a faculty of architecture. Students do not often have the opportunity to practise the interdisciplinary collaboration that is so important later on. The focus here is more on civil engineering, i.e. road, bridge, and hydraulic engineering, and less on structural engineering, where civil engineers work together with architects to plan buildings. My long-standing membership in the International Association for Bridge and Structural Engineering (IABSE/IVBH) is paying off. I had also previously met colleagues from here at IABSE symposia, and they now welcomed me warmly. I have known Jan Wium, the professor of structural concrete and construction management with the square-edged face, whose English always has a hint of Afrikaans, for over 20 years. He has to retire in 2024, but that is too soon for him. He will, therefore, move to Zambia with his wife in 2024 to set up a civil engineering degree programme at Northrise University in Ndola. "Develop Africa from within" is his credo. Here in Stellenbosch, Jan shows me with justified pride the civil engineers' new office building and the impressively large water laboratory for experimental simulation of watercourses Fig. 3. A 20-metre-long scale model of a reservoir with a dam, hydroelectric power station, and lakebed made of fine gravel is currently being built there. The scientists will slowly fill the model with water and then drain the lake at the power plant. In this way, they want to investigate how the water currents change the lakebed. I repay him for the informative tour with a lecture on conceptual and structural design and our teaching and research at the TU Berlin.

A few days later, I experience hospitality at the University of Pretoria, where I am greeted by a robot dog controlled by a doctoral student at the newly inaugurated Engineering 4.0 laboratory, where various engineering disciplines conduct joint research Fig. 4. Everything I see here, from the laboratory to the lecture hall, is enviably well equipped and allows for research and teaching at the highest level. Professor Elsabe Kearsley, former president of the South African Institution of Civil Engineering (SAICE) and the South African Academy of Engineering (SAAE), has been teaching here since 1990 and takes time out for me. She introduces me to her colleagues, recommends books on civil engineering in South Africa, and kindly agrees to contribute a text to this book. Her young and energetic civil engineering colleague Sarah Skorpen, who picked us up from the airport and organised my lectures here, insists on inviting us to her home for a braai, a typical South African barbecue in the garden with lots of meat.

At the third engineering university I have the privilege of visiting, the venerable University of Cape Town, founded in 1928, I meet Professor Alphose Zingoni, co-organiser of the international SEMC conference on structural engineering, mechanics, and computation, held every three years. He immediately plans to include me in the next event. He has made a name for himself particularly in the field of thin-walled shell structures. Alphose, whom I have known for years, is about my age and originally comes from Zimbabwe. He studied and obtained his doctorate at Imperial College London and has been teaching here in Cape Town as a professor since 2002.

3

4

5

Figs. 3–5 The hydraulic engineering laboratory at Stellenbosch University, a robot dog at the University of Pretoria, and the concrete laboratory at the University of Cape Town

Concrete laboratories Fig. 5 are standard equipment at any civil engineering institute, and at all three universities I find workshops and equipment, such as concrete creep stands that we would also like to have at TU Berlin. When I ask about the best civil engineering education in the country, the following universities are regularly mentioned:

→ 1. Stellenbosch University
→ 2. University of Cape Town
→ 3. University of Pretoria
→ 4. University of the Witwatersrand
→ 5. University of Johannesburg
→ 6. University of KwaZulu-Natal

The discussions at the universities resulted in the following impressions:

→ As at most of the universities I visited later on, the education system follows the British-American model with few full professors and many lecturers, often civil engineers with master's degrees.
→ Our system, with its mainly full professors and research assistants, is the envy of others. The practical experience of German professors is also considered desirable.
→ South African universities are attractive to students and doctoral candidates from other African countries. For them, a doctorate in South Africa is often a springboard to a professorship in their home country. In contrast, universities in South Africa tend to have more contacts with Europe than with other universities on the continent.
→ Teaching and research focus on the traditional areas of civil engineering, such as geotechnical engineering, mining, road construction, and hydraulic engineering.
→ Bridge construction is rarely taught and the subject of *Baukultur* ('building culture' is the closest translation but I will continue to use *Baukultur* in this book) is not covered. The responsibility of civil engineers for issues of sustainability and climate change is only slowly finding its way into teaching and research.
→ South Africa does not have a system of apprenticeships as we know it in Germany. Bricklayers and concrete workers are trained through learning by doing on the construction site.

Day 14, 28 March – Cape Town 2

Despite all the quotas set by the government, it seems to be taking a long time for the ethnic realities of the country to be reflected in the bridge engineering profession. My predominantly white colleagues from the city's planning offices take time for me, and I have the opportunity to talk to John Anderson from SMEC, who designed the rotating, cable-supported Clocktower Pedestrian Bridge at Cape Town harbour Fig. 6. As with Calatrava's Alamillo Bridge in Seville, this slender bridge with its central steel pylon and four stay cables lacks backstay cables, but this is forgivable given the low pedestrian loads here. I meet Henry Fagan, the civil engineer who was also involved in the design of the German ambassador's residence in Cape Town, in the Kirstenbosch Botanical Gardens. There, the elderly gentleman in a leather slouch hat, with a youthful enthusiasm for *Baukultur*, shows me his Boomslang Bridge, which winds elegantly around the treetops and offers fantastic views of the garden and Table Mountain Fig. 7. I also have the opportunity to speak with Edwin Kruger, who shaped road bridge construction in South Africa for many years on behalf of SANRAL (South African National Roads Agency Limited). He has long since reached retirement age, but civil engineering still holds a fascination for him. I will see him and John again at the Msikaba Bridge construction site. The conversations with these experienced colleagues give me the following insights into the practice of civil engineering in South Africa:

- → South Africa is one of the few countries in Africa with its own established steel factories and national iron ore mining. The first steelworks were built at the beginning of the twentieth century, and in 1928 the semi-state-owned industrial company ISCOR, the Iron and Steel Corporation Ltd., was founded to make the country independent of steel imports.
- → South Africa's strong mining industry supplies building materials. For example, spoil from South Africa's many mines is used for dam construction.
- → However, as everywhere in Africa, the predominant material is concrete. Cement and aggregates are sourced within the country.
- → The closure of coal-fired power stations and coal mines will reduce the availability of fly ash, a waste product used as an additive in cement to reduce its carbon footprint. South Africa is, therefore, also looking for substitutes for less environmentally harmful concrete.
- → Winters in South Africa are snow-free. The corrosion of reinforcement caused by de-icing salt, which is such a problem for German bridges, is therefore not an issue here.
- → Due to high unemployment, labour-intensive construction methods are chosen on building sites. This is also one reason why in-situ concrete is used more than precast elements in the country.
- → The long distances to the often remote and difficult-to-access construction sites for infrastructure projects make equipment transport costly and influence the choice of structures and construction methods more than in Germany.
- → The government is trying to protect itself from excessive foreign influence by requiring the use of domestically produced steel and cement for public buildings. However, the state has neglected housing construction in recent years. As everywhere in Africa, there is a shortage of affordable housing.
- → As in Germany, engineering firms are undergoing a process of consolidation. For example, the engineering firm VKE, founded in the 1950s, was acquired in 2012 by the Australian SMEC Group, which in turn was incorporated into the even larger Surbana Jurong Group from Singapore in 2016.

6

7

Figs. 6–7 The Clocktower footbridge and the Kirstenbosch treetop walk in Cape Town with Henry Fagan

8

9

Figs. 8–9 Overlander exhibition near Stellenbosch with off-road trailer and South African pickup truck with travel attachment at Crocodile Safari Camp in Maun, Botswana

Day 25, 8 April – Mthatha

In the meantime, we – my friend and 'co-pilot' Massimo Canossa from Bologna has also been here for two weeks – are preparing for the journey north: another chain with a padlock to secure the second spare tyre to the roof and red (left) and white (right) reflectors for the bumpers, because we read that you can get a fine in Zambia if you don't have them. We buy provisions for the trip in Cape Town's supermarkets. We are surprised to find the same range of products as at home; there are fruits and vegetables from avocados to zucchini. Massimo is a vegetarian, which means I will be eating healthier than usual for a few months, and because he is an agricultural engineer who has monitored development aid projects around the world, I will learn a lot on this trip about the sense and nonsense of the agricultural activities we will see.

From the seventeenth century onwards, Dutch settlers colonised the interior of the country. Their Afrikaans-speaking descendants, whose surnames can be easily distinguished from those of the British who arrived later, have retained this settler spirit in the form of a love of travel. Everywhere in southern Africa, we encounter these South African tourists travelling in incredibly well-equipped recreational vehicles. And everywhere in South Africa, you can find workshops and exhibitions for 'overlanders' with equipment that makes the heart of every Transformers fan beat faster. An inconspicuous little trailer can be transformed into a high tent with an outdoor shower and kitchen in no time at all, and a living module can be mounted on a farmer's pickup truck in minutes, turning it into a comfortable motorhome for difficult terrain Figs. 8–9. Kobus du Plessis, a hydraulic engineer in Stellenbosch and a fellow professor, even drove his caravan with his family over hills and dales to Kenya and back again without incident. Custom-made cars like our Land Cruiser, built for long-distance travel only and, therefore, unsuitable for everyday use, are regarded in southern Africa more as rental cars for tourists.

We set off on our journey and are cautious. On the way from the airport to Cape Town, we saw signs urging drivers not to stop on the motorway under any circumstances and to call the police immediately in the event of an accident. Caution is also advised near townships, places in no man's land such as the Cape Flats near Cape Town, where the black population was forcibly resettled in the 1950s. South Africa still suffers today from the consequences of these apartheid crimes and the hopelessness of these 'cities'. Children left to fend for themselves find drugs and supposed father figures in the local gangs, and criminal initiation rites are part of rising through the ranks within the gang. The colleague who explains these connections to me ends his account with a warning: "Here you get killed, not only robbed."

It is the rainy season, and as we enter Mthatha, a uniformed officer stops us. He informs us that the road through the city is flooded, points to a car that has just turned off, and recommends that we follow it. At least we are able to continue, and we accept that we will have to pay a fee for the private road we now have to use. The fluent English-speaking, courteous driver of the car in front of us also shows us where the somewhat difficult-to-find ticket machine is located. It is only when I see my PIN written out on the screen as I type it in and the card gets stuck in the machine while more helpers come closer and closer that it dawns on me that something is not right here. I stop the process, raise my voice, and call the police (the emergency number is 112). Suddenly, the card comes out of the machine again and all the helpers disappear as quickly as they appeared. I blocked the card within 15 minutes, but our helpers have already withdrawn 2,000 euros. With our heads clear again and alone, we realise that the barrier to the private road we passed belongs to a hospital, the uniformed man from earlier has disappeared, and there is no flooding anywhere in the city. We realise that we have been extremely lucky. Had we been lured to this lonely car park by less scrupulous people instead of professionals, it could have ended really badly for us. In Mthatha, we were taught a harsh lesson about uniformed men and gullibility, which continued to have an effect until the end of the trip.

Day 28, 11 April – Msikaba

On the very day of our visit to the construction site, the Eastern Cape is hit by the heaviest rainfall since records began in 1921. The subtropical depression Issa claims over 400 lives and causes damage amounting to 1.5 billion US dollars. [2] Small streams become raging torrents and pour into deep gorges as thundering earthy waterfalls in this mountainous rainforest region. The water-filled potholes on the unpaved tracks do not reveal their depth, and on every incline we worry whether our studded tyres will cope with the road surface, which has turned into brown mud.

Despite the state of emergency, we manage to reach the Msikaba Bridge construction site near Lusikisiki safely in our now earth-coloured car, covered in mud up to the windows. With a span of 580 metres, it will be Africa's longest cable-stayed bridge. South Africa already boasts world-famous bridge structures such as the Bloukrans Bridge along the Garden Route between Cape Town and Durban. Bridge construction is the supreme discipline of civil engineers, but the architects at Dissing+Weitling, with their decades of experience in bridge design, are welcome sparring partners in the bridge designers' club. Together with engineers from Parsons and SMEC, they have designed a bridge here in the mountains of KwaZulu-Natal, between sprawling tea plantations and untouched nature, which will enrich South African *Baukultur* and noticeably shorten the journey along the N2 Wild Coast Toll Road to Durban.

Between two slender, 127-metre-high concrete pylons, two cable planes will support a 22.8-metre-wide composite deck for four lanes. The view of the 200-metre-deep valley with its completely untouched lush vegetation, where leopards are said to still roam, justifies the large span. So far, only the mast bases on both sides of the valley are visible. Behind them are huge excavation pits for the concrete abutments in which the backstay cables will be anchored. The valley is currently still crossed by a cable car purchased in Austria. In this remote and inaccessible area, the journey to the other side would take several hours by car Figs. 10–11.

10

11

Figs. 10–11 Construction site of the Msikaba Bridge and view of the valley it will cross

The tireless and always cheerful John Anderson, a Scotsman who emigrated to South Africa many years ago and recently became CEO of all SMEC branches in Africa, took the time to bring me here. With rain pattering on the corrugated iron roofs of the construction site camp, which has many containers for offices, storage, and a canteen, I have the opportunity to give a presentation on our cable-stayed bridges to the engineers on site and hear about their experiences here. Open exchange among colleagues is also cultivated here and I feel immediately accepted by my peers.

The bridge is scheduled for completion at the end of 2025. However, it has been postponed to the next year for several years now. This is also due to South Africa's protectionist system. Quite rightly, the country wants to limit the influence of foreign companies and investors and stipulates a minimum 50 per cent national participation in infrastructure projects. In addition, in an attempt to eliminate the consequences of apartheid, the local economy must be involved so that local expertise can be developed. The two construction companies building the bridge, Concor and MECSA, are 'both 100 per cent black-owned Grade 9CE South African construction companies', and 70 per cent of the workers come from the surrounding area. [3] The steel wires for the bridge cables are also set to come from South Africa. This condition poses a serious challenge for the French company Freyssinet, which is responsible for the cables.

Cables for cable-stayed bridges require the highest fatigue resistance and processing quality, for which South Africa is not prepared, and the local construction industry and workers must be trained at great expense. It is, therefore, possible that the inauguration date will be postponed again. However, if this means that South Africa gains more experience in bridge construction, then the costs will have been worthwhile.

Civil Engineering in South Africa
Prof. Dr. Elsabe Kearsley,
University of Pretoria, South Africa, April 2024

'Since 1890, there have been various legal regulations in South Africa governing the licensing of engineers. The Engineering Profession Act (EPA) of 2000 led to the establishment of the Engineering Council of South Africa (ECSA), a state institution responsible not only for the registration of engineers, but also for the accreditation of all engineering training programmes in South Africa. The prerequisite for registration with the ECSA is a recognised engineering qualification, followed by at least three years of on-the-job training, during which the trainee works under the supervision of a registered engineer. Currently, more than 26,000 engineering professionals are registered with the ECSA. South African universities have been offering degrees in civil engineering for more than a century, and there are currently 18 public universities that train engineers, technologists, and technicians in various engineering disciplines. The ECSA is a co-signatory to international agreements, which means that South African engineering degrees are recognised internationally.

Before the construction of the Suez Canal, the Cape of Good Hope at the southern tip of Africa was the hub of the maritime trade route between Europe and the Far East. Portuguese, Dutch, and British trading companies established trading posts on the Cape coast to supply themselves with water and food for the long voyages. Settlers from Great Britain, the Netherlands, France, Germany, Malaysia, India, and China encountered indigenous nomadic tribes who roamed Africa in search of areas with good rainfall and sufficient food. This mixing of different cultures led to the unique works of architecture and engineering in South Africa.

The indigenous African tribes used local materials such as reeds, grass, clay, and wood to construct buildings that provided protection from the weather. Although these were not very durable, they were well suited to the hot, dry climate due to their good ventilation and thermal insulation. Thatched round huts can still be found in many parts of South Africa today, and modern city dwellers often use traditional construction methods for their homes.

One of the first large permanent structures in South Africa was built in 1652: a fort made of clay and wood, erected by the Dutch East India Company (VOC) during the founding of the settlement at the Cape. In the following years, several mountain passes were opened up to allow access to the interior of the country. Many of these early mountain passes are still in use today.

The discovery and extraction of mineral resources such as diamonds, gold, platinum, coal and uranium drove development in large parts of South Africa and resulted in a well-developed road network that today comprises 23,559 kilometres of national roads.

Dutch, British, French, and German architects, builders, and engineers combined their expertise acquired in their home countries with local traditions to create unique buildings such as Cape Dutch gabled houses, thatched Georgian buildings, and large sandstone churches in many small towns and settlements. The sandstone Union Buildings in Pretoria are probably South Africa's best-known building complex, a true architectural and engineering masterpiece designed by Sir Herbert Baker. There are also picturesque decorated buildings, such as those of the Southern Ndebele ethnic group, adorned with bright geometric patterns, and the colourful houses in Bo-Kaap in Cape Town.

With annual rainfall of only 464 millimetres, South Africa is a water-scarce country, and engineers had to design complex water supply systems with large dams, tunnels, and pumping stations to supply water to the Cape Town metropolitan area and the province of Gauteng, where the capital Johannesburg is located and where the majority of the population lives. Most of the water for Gauteng comes from the Lesotho Highlands Water Project, under which South Africa will obtain approximately 1,260 million cubic metres of water annually, and Lesotho will use the large dams to generate hydroelectric power.

The 2010 FIFA World Cup gave the construction sector in South Africa a significant boost in the run-up to the event, as numerous stadiums were built, including the iconic Soccer City Stadium in Johannesburg, Greenpoint Stadium in Cape Town, and Moses Madiba Stadium in Durban. The road and rail network in Gauteng was also significantly expanded, as exemplified by the Gauteng Freeway Improvement Project (GFIP) and the Gautrain regional rail network. Most of the planning and construction of these major infrastructure projects was carried out by local engineers and construction companies. This created a large number of jobs in a country where unemployment is one of the biggest challenges.

South Africa is a melting pot of cultures with a rich and diverse engineering heritage. This country is undoubtedly the perfect starting point for a journey to study construction on the African continent!'

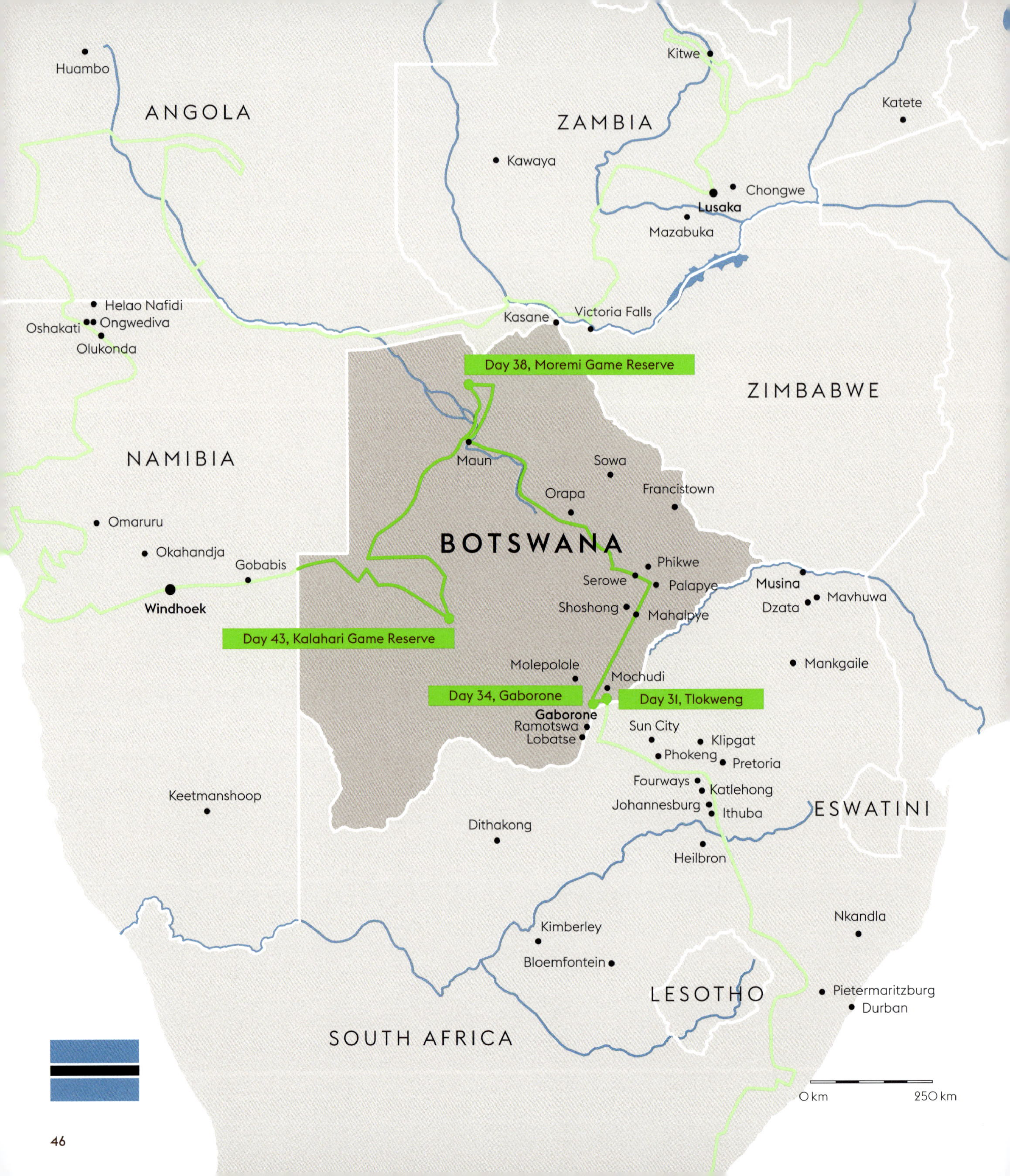

Huambo
ANGOLA
ZAMBIA
Kitwe
Katete
Kawaya
Chongwe
Lusaka
Mazabuka
Helao Nafidi
Oshakati
Ongwediva
Olukonda
Kasane
Victoria Falls
Day 38, Moremi Game Reserve
ZIMBABWE
NAMIBIA
Maun
Sowa
Orapa
Francistown
Omaruru
BOTSWANA
Okahandja
Gobabis
Phikwe
Serowe
Palapye
Musina
Mavhuwa
Dzata
Windhoek
Shoshong
Mahalpye
Day 43, Kalahari Game Reserve
Mankgaile
Molepolole
Mochudi
Day 34, Gaborone
Day 31, Tlokweng
Gaborone
Ramotswa
Lobatse
Sun City
Klipgat
Phokeng
Pretoria
Fourways
Katlehong
Keetmanshoop
Johannesburg
Ithuba
ESWATINI
Dithakong
Heilbron
Nkandla
Kimberley
Bloemfontein
LESOTHO
Pietermaritzburg
Durban
SOUTH AFRICA
0 km
250 km

Botswana
14–28 April 2022

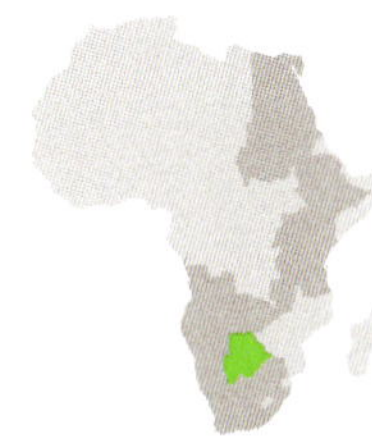

Botswana has been fortunate. Immediately after the British granted independence to this country, which they considered economically uninteresting, diamonds were discovered here. Today these precious stones account for more than 70 per cent of Botswana's exports and, thanks to good governance, have brought the country steady economic growth ever since. Tourism provides additional income. The sparsely populated country, with only 2.3 million inhabitants, is plagued by AIDS. Approximately 20.3 per cent of 15- to 49-year-olds are infected with HIV. [1]

Day 31, 14 April – Tlokweng

At the Tlokweng border control point, we learn about the border formalities for car travellers in Africa for the first time. A Covid test accompanied by a temperature measurement is standard, and we wait patiently at immigration and customs until our visa and Carnet de Passages are stamped. Depending on the country, we purchase a road permit and local liability insurance before paying taxes such as city or carbon tax. To ensure that everything is above board when making such payments, we have to deposit the respective amount at a bank and present the receipt at the appropriate counter. With each border crossing, we become more relaxed, learn to fend off the many 'fixers' who want to help, and find the responsible officials increasingly quickly. Before the barrier finally goes up, we submit all our new papers and stamps to an official for a final check. Botswana makes it easy for us to enter the country. Immigration and customs are dealt with quickly, though Covid-related checks take us an hour because the EU Covid Certificate is only accepted after some discussion.

Peter Reger from the German Society for International Cooperation (GIZ) is already waiting for us on the main road with his bicycle. We have reached Gaborone, the capital of Botswana, in just two hours, and now he is cycling ahead of us on the dusty side road so that we don't get lost. On a nameless street, hidden behind walls, he has rented a spacious villa for his stay here – one we never would have found without a guide. The sturdy lattice doors are opened, and from the veranda we look out onto green lawns, palm trees, and a barbecue area. The house itself has easy-to-clean stone floors and is furnished in a spartan manner: after all, you're not staying forever. Gaborone is also home to the Southern African Development Community (SADC), an association of 16 southern African states that hope to one day achieve peaceful and free coexistence, as in the European Union. For this reason, and due to the country's stability, virtually all Western development agencies and NGOs are based here. In Botswana, Peter and his colleague Helmut Riethmüller are currently trying to integrate apprenticeships based on the German model into the training landscape and pave the way for a large training centre. The tall building they have in mind is to look like a large baobab tree. I contribute sketches for the possible supporting structure. The rounded trunk accommodates all the rooms and is to become a perforated, freely formed but efficient external reinforced concrete cylinder, stiffened by the floor plates. The branching limbs are formed by steel cantilevers that

Fig. I Bronze statues of the founding fathers of Botswana in Gaborone

can support viewing platforms and solar panels and symbolise the leaves. This could be a rather unconventional but sensible construction, and I would love to start planning right away, but let's take it slowly: first, potential clients need to be enthusiastic about the idea. Hopefully, these two idealists will have the staying power that will be needed for this project . In any case, Peter is the epitome of calm – a long expat life with development projects all over the world probably has something to do with that. He is currently plagued by a different problem: he wants to cook for us and can't find any onions. Botswana is trying to become independent of imports from South Africa, and a sudden ban on imports of agricultural products has led to an onion shortage throughout the country.

The city can be explored quickly. Where thatched-roofed huts once stood, glass-fronted skyscrapers are now being built in the central business district, cooled by electricity from coal-fired power stations. Renewable energies from wind and solar power plants currently account for less than I per cent of electricity generation. The main attraction in the centre is a set of three bronze statues, each measuring over 5 metres in height, made in North Korea and depicting the founding fathers of the country, the Tswana rulers Khama III, Sebele I, and Bathoen I Fig. I. In Namibia, Angola, and Ethiopia, we will also encounter impressive monuments that come from the 'brother people' in the Far East.

Day 34, 17 April – Gaborone

Botswana has two new universities: Botswana University in Gaborone, founded in 1982, and the Botswana International University of Science and Technology (BIUST) in Palapye, which only opened its doors to students in 2011. Both train civil engineers. I am visiting the former. Although nobody is here due to the Easter holidays, the modern building of the Faculty of Engineering and Technology, constructed from brown-coloured reinforced concrete and brickwork, is open and I can wander around undisturbed. Someone has left a pile of marked structural analysis tests lying around, so I can check at my leisure that Nandi M. has solved his truss calculation excellently. Figs. 2–3.

We stock up on crispy fresh sourdough bread at the Avani Hotel and set off north to the national parks. The country's standout attractions are its two wildlife paradises: the green Okavango Delta and the sandy Central Kalahari Game Reserve.

Any car can cover the route from Gaborone past Serowe and Orapa to Maun in the north. Travellers will not miss petrol stations or the Debonairs pizza restaurants we have come to appreciate in South Africa. The only thing to watch out for is speed cameras, as fines are non-negotiable in Botswana. We also have to get used to the African plague of artificial road bumps made of concrete or asphalt, known as 'speed humps', 'speed bumps', 'speed breakers', 'lying policemen', or 'dead policemen', which are used to limit speed. They vary greatly in height and often cannot be driven over at more than 20 km/h without causing trouble. Because they are not always marked and can take you by surprise outside of built-up areas, they can cause serious damage to your car and slow down the movement of people and goods. For this reason, too, we avoid driving at night and, if possible, at dusk. This is also because grey animals such as donkeys and elephants, which are often on the roads, are virtually impossible to see at these times. Figs. 2–3

2

3

Figs. 2–3 The Department of Civil Engineering at Botswana University in Gaborone and a view inside the building

Day 38, 21 April – Moremi Game Reserve

In the Okavango Delta, the largest inland delta in the world, the Cubango River from Angola seeps away and forms a landscape of lakes and swamps with unparalleled fauna and flora. We cannot stop marvelling at it. Elephants, hippos, and lions take turns appearing, and in our 'hunting fever' we have to be careful not to overburden our car in the swampy terrain. We pull a tourist couple out of the mud; they got stuck because no one had explained to them that the rental Toyota runs in two-wheel drive and has to be manually switched to four-wheel drive by hand if needed. We are proud of ourselves Fig. 4. However, you shouldn't imagine this kind of trip as an expedition. The national parks in Africa are open-air zoos, some of them huge and covering over 10,000 square kilometres, where visitors have to pay an entrance fee and must remain on designated paths, which vary in accessibility depending on the season. The slightly smaller Moremi Game Reserve spans 5,000 square kilometres and is still twice the size of the German state of Saarland. I take a close look at the very simple wooden bridges in the park before we cross them with our vehicle, which weighs almost three tonnes Fig. 5.

Day 43, 26 April – Central Kalahari Game Reserve

In the Central Kalahari Game Reserve, which covers just over 50,000 square kilometres, we manage to secure one of the few 'campsites'. We are given the GPS coordinates for our spot at Xade II, and we find a site in the bush with a small fire pit. We have brought firewood and water with us, and we dig our own toilet. The nearest campsite of this kind is 50 kilometres away, and we enjoy the unique clarity of the Milky Way in this solitude, with night-time temperatures of a pleasant 15° Celsius. The fire successfully protects us from wild animals, and the next day we cross the vast grasslands full of acacia trees at 35° Celsius on deep sand tracks, which we share with elephants Figs. 6–7.

The two most interesting bridges in Botswana, both cable-stayed, are located outside the national parks in the far north of the country. The Okavango River Bridge in Mohembo, with a main span of 200 metres and a total length of 1,161 metres, carries a two-lane superstructure. The Hamburg-based engineering firm GRASSL, which designed the bridge, writes: "It crosses the Okavango and serves to open up the eastern part of the Okavango Delta. The steel pylons of the cable-stayed bridge symbolise the 'gateway to Africa' through their design. The bridge deck is formed by a reinforced concrete composite section." [2] The bridge will be inaugurated in July 2022 – we drive past the construction site without a clue. It is only in Germany that Martin Grassl tells me about 'his' bridge. What a missed opportunity! However, the Kazungula Bridge over the Zambezi River, located about 400 kilometres to the west, is significantly larger and more important for the country. This five-span extradosed bridge, over 900 metres long, opened in May 2021 and not only connects Botswana with Zambia, but also facilitates the movement of goods between South Africa and the Congo (DRC). (There will be more about this bridge in Zambia on day 66.) Concrete is also the material of choice for building construction in Botswana. Houses are often built with solid concrete blocks, but there seems to be a slow shift towards lightweight hollow blocks, which consume fewer resources and provide better thermal insulation.

4

5

6

Figs. 4–5 Rescue in the Moremi Game Reserve (finally, our winch is needed!) and bridge crossing in the Okavango Delta

7

Figs. 6–7 Even elephants prefer the road to the path through the acacia scrub of the Kalahari

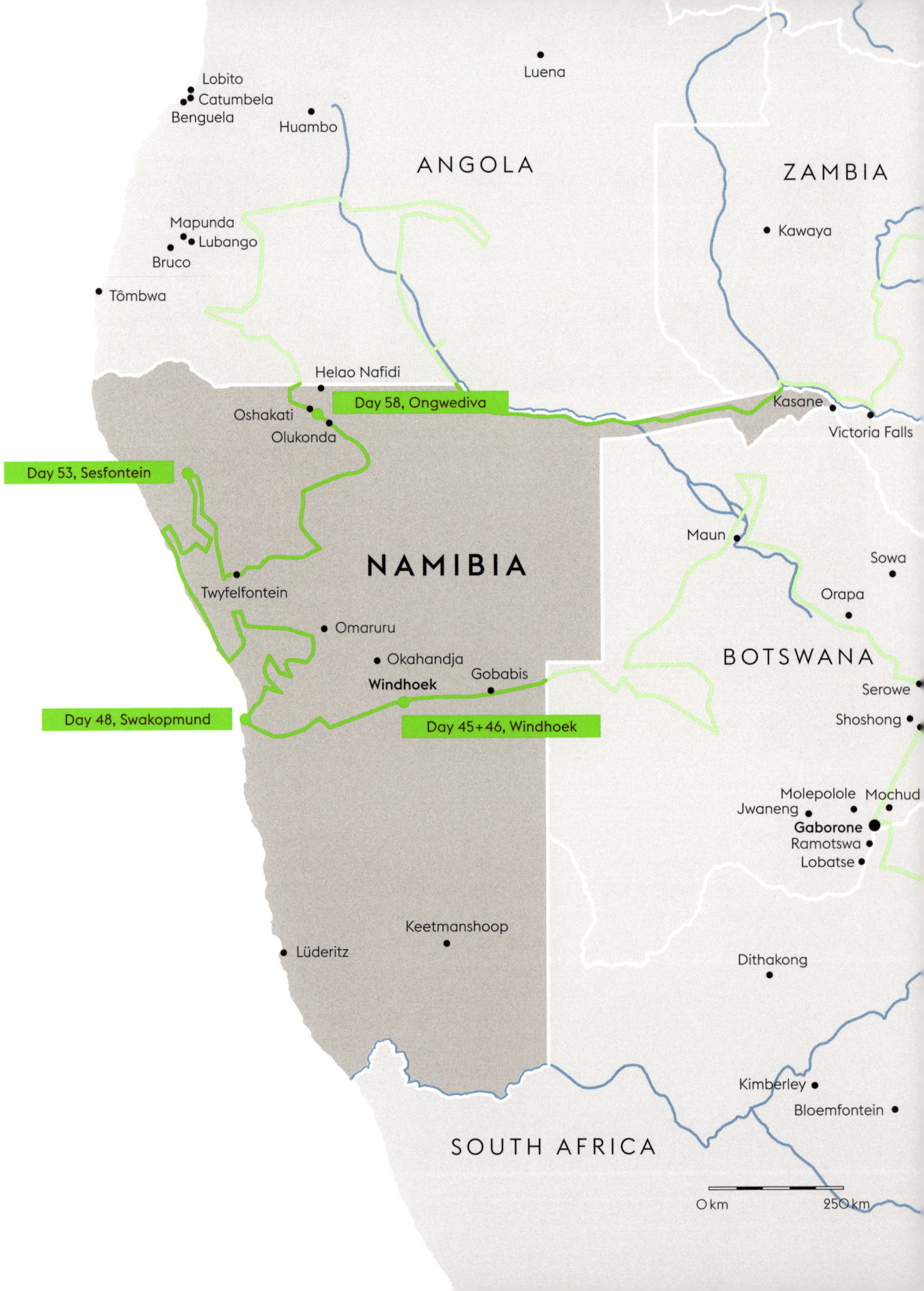
ANGOLA
ZAMBIA
NAMIBIA
BOTSWANA
SOUTH AFRICA
Luena
Lobito
Catumbela
Benguela
Huambo
Mapunda
Lubango
Bruco
Tômbwa
Kawaya
Helao Nafidi
Day 58, Ongwediva
Oshakati
Olukonda
Kasane
Victoria Falls
Day 53, Sesfontein
Maun
Sowa
Twyfelfontein
Orapa
Omaruru
Okahandja
Gobabis
Windhoek
Serowe
Day 48, Swakopmund
Day 45+46, Windhoek
Shoshong
Molepolole
Mochud
Jwaneng
Gaborone
Ramotswa
Lobatse
Keetmanshoop
Lüderitz
Dithakong
Kimberley
Bloemfontein
0 km
250 km

Namibia
28 April–12 May from Botswana and 17–19 May 2022 from Angola

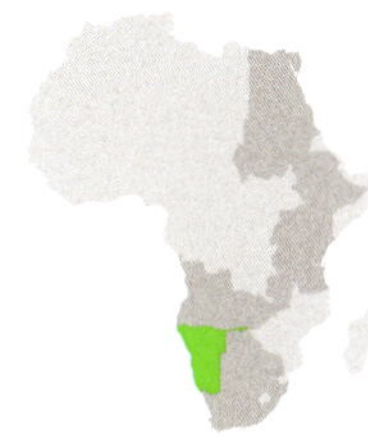

Crossing the border in your own car in Africa remains exciting and, for me as a European, unfamiliar because you have to venture into the completely unknown for a short time and be prepared for all kinds of surprises. Do I take the border crossing on the side road where there may be no computers to check my e-visa and the border guards' lunch breaks are longer? Or should I take the main road where there may be long queues and all kinds of 'helpers' who might take advantage of my ignorance? Do I have all the right papers? In what order do I have to queue and where? And are the fees payable in local currency, dollars, or by card? There is only one border crossing between the Kalahari in Botswana and Windhoek in Namibia. We have also learned that there is little bureaucracy to contend with here because Namibia is part of the South African Customs Union (SACU) along with South Africa and Botswana. Crossing the border is a breeze; the tension beforehand was unnecessary.

Day 45, 28 April – Windhoek

In bright sunshine, we drive straight west through a deserted steppe and bush landscape to the capital Windhoek, which is beautifully situated in the centre of the country and has a population of just under half a million. With only around three inhabitants per square kilometre, Namibia is even more sparsely populated than Botswana. With about 19 kilometres of road per inhabitant (more than three times the German figures), it is obvious that there cannot be enough money to asphalt all the roads in Namibia. As a civil engineer, you have to be able to design dirt roads here, and indeed, Namibia knows a thing or two about planning unpaved roads. Driving on these well-drained and mostly pothole-free roads is surprisingly pleasant. However, traction is poor on gravel and crushed stone, and we do not drive faster than 70 km/h so that we do not lose control of the car when braking. When driving down winding roads, we slow down even more, as wrecked cars on the side of the road serve as a warning.

You can also build roads with salt! The C34 runs along the coast from Henties Bay to Terrace Bay on the Skeleton Coast and offers these exotic road surfaces. These salt roads, probably the only ones of their kind in the world, were built in the 1950s and 1960s. [I] Like morning mist, the lonely, brownish-dark track is covered with whitish veils. It is the Atlantic wind that blows light, fine sand from the Namib Desert over the salt. Despite the high level of driving comfort on the perfectly flat and grippy salt surface, we are glad when we are back on less corrosive road surfaces Fig. I, p. 60.

2

3

Figs. 2–3 Solar panels at the lodge near Cape Cross, where Diogo Cão became the first European to land in 1486, and the remains of one of the many shipwrecks on the Skeleton Coast

Sun and wind are abundant on Namibia's long Atlantic coast. They are ideal for stand-alone energy supply solutions outside urban areas, i.e. Windhoek. In fact, in the few settlements on the Skeleton Coast, which are mainly visited by sport fishermen from South Africa, we see solar panels as an alternative to diesel generators Fig. 2. The dry desert wind blowing against the breaking Atlantic waves here shows its strength by sending white spray metres into the air Fig. 3. However, I cannot see any wind turbines along this shipwreck-strewn coastline; renewable energies currently contribute less than 1 per cent to the country's electricity supply. Namibia wants to change that. The huge Hyphen project in the Namib Desert near Lüderitz in the south of the country, which launched in 2022, is to be brought up to a capacity of 7 gigawatts – equivalent to seven nuclear power plants – using a mix of photovoltaics and wind power. With planned electricity production costs of 2 ct/kWh, hydrogen is also to be produced, which will benefit the country and can also be exported. The company Enertrag from the Uckermark region is involved in this Namibian-German project. [2]

Next to the B6 road to Windhoek International Airport, the historic railway bridge bears witness to railway bridge construction before independence. I rarely see steel truss bridges of this type, prefabricated in England or later in South Africa Fig. 4. The most common 'bridges' in the countryside are culverts, which are elevated rainwater drains. They consist of precast concrete modules or segments with a span of two to ten metres, which allow road traffic to pass over an otherwise dry riverbed at a low height even during the rainy season. We see structures like this all over the country, and Roland Dishena, Chairman of the Namibian CAT Group, tells me on LinkedIn that the Namibian precast companies Vula Namibia and Rocla Namibia have sold an impressive 15,000 such culvert elements over the past six years Figs. 5–8.

4

5

6

7

8

Figs. 4–8 One of the rare historic railway bridges (4), culverts (5), small bridges for the rainy season in Namibia (6–7), and an example from Zambia (8)

Day 46, 29 April – Windhoek

The Namibian construction industry is currently struggling. This is particularly true for building construction, even though affordable housing is urgently needed for the growing population. As there is not much work available, mostly small local engineering firms and construction companies have to accept any job they can get. However, due to a lack of skilled workers, this has an impact on the quality of construction, especially in steel construction. In addition, competition from Chinese construction companies is noticeable, especially for large contracts. In the city, near the former apartheid settlement of Katutura, I have already seen the new headquarters of the ruling party, the South-West Africa People's Organisation (SWAPO), which is currently being built in an opulent Chinese style.

As in South Africa, concrete is the main building material in Namibia, and even masonry is often made of solid concrete. The cement can be supplied by two Namibian factories. Steel, aluminium, and wood are imported from South Africa, where many of Namibia's civil engineers studied. Even the north of the country is supplied from there because there are hardly any established links with neighbouring Angola.

There are no snow or earthquake loads in Namibia. Shouldn't it be easy to build with steel? Not everywhere! The Atlantic Ocean drives salt spray inland. There are serious corrosion problems up to 50 kilometres from the coast. This leads to spalling even in concrete structures due to reinforcement corrosion. Concrete drinking water reservoirs, which this water-scarce country depends on, are particularly at risk here.

Are civil engineers in Namibia concerned with issues of sustainable construction, local materials, and traditional building methods? Here I am referred to Namibian architect Nina Maritz, who also works on sustainability issues and whom I have already come across in an interview. [4] She has built in surprising ways throughout the country, collaborating with local craftsmen and using local materials. In the interview, she describes her buildings, with walls made of rammed earth or gabions made of local stone, and the obstacles that slow down sustainable construction: "The main obstacle to the use of alternative materials lies in a catch-22 situation of conservative financial institutions (banks and building societies), who refuse to issue loans for any alternatives unless approved by the local authorities, and local authorities refusing to approve alternatives unless they qualify for a loan. A secondary obstacle to using alternative materials is the public perception that the only 'proper' housing is built from typical red earthen brick."

In German-speaking countries, Anna Heringer and Martin Rauch are known for building with clay and earth, and I suspect that they have to contend with similar problems. In any case, it sounds like the situation here in Germany. Although there are enough solutions, sustainable construction projects repeatedly fail due to a series of obstacles and resistance. Hesitant builders, overly cautious planners, checking engineers, or insurance companies can be just as much to blame for this as excessive regulation, a lack of acceptance, and comfort zones that people are reluctant to leave. That is why we have set up the interdisciplinary working group *Anders Bauen* (*Building Differently*) at the Berlin-Brandenburg Academy of Sciences and Humanities. There, civil engineers will work together with representatives from other disciplines, such as sociologists, psychologists, and legal scholars to investigate why such resistance arises and how it can be overcome. [5]

Day 48, 1 May – Swakopmund

Coming from the cloudless desert, we head towards a huge white wall. Clouds pile up along the Atlantic coast, enveloping Swakopmund in thick fog in the early hours of the morning. As it clears, we find ourselves in a former German colonial town where much sounds familiar. We spend the night at the Princess Rupprecht Residence, eat at the Deutsches Haus restaurant, and visit the Woermannhaus.

Because the British, 70 km to the south, claimed the well-protected harbour in Walvis Bay for themselves, the Germans had to build a barely usable harbour in Swakopmund, which repeatedly silted up, so that their settlers could reach what was then German South West Africa under the protection of the German *Schutztruppe*. In order to be able to unload larger ships, a wooden jetty was first built, but it was damaged by waves and shipworms brought in by the Germans themselves. From 1911 onwards, serious work began, and a 20-metre-wide and 650-metre-long steel landing stage with large cranes was to be built out into the sea beyond the surf. The companies Flender AG and Grün & Bilfinger were commissioned to drive concrete-filled and galvanised steel pipes into the ground for the pile foundations. After slow progress, the project was finally abandoned shortly after the outbreak of war in 1914. The recently published, extensively researched anthology *Deutschkoloniale Baukulturen ('German Colonial Building Cultures')*, edited by Michael Falser, features a photo of this impressive large-scale structure [6]. As early as 1903, a railway line with telephone masts and telegraph lines was built from Swakopmund into the interior of the country. At the time of its completion in 1906, the Otavi Railway was the longest narrow-gauge railway in the world, with 567 kilometres of track. 110 bridges were built for it, including the Khan Bridge near Usakos, which is over 100 metres in length and whose superstructure was formed by five single-span girders, each with a span of 20 metres. "Rolled iron girders were used, prefabricated from eight different spans and riveted together on site." [6] The truss girders of the Bösebrücke and Swinemünderbrücke bridges in Berlin were built around this time. So too were the famous AEG turbine hall in Berlin Moabit by Peter Behrens and the even larger Peter Behrens Hall in Wedding, where AEG assembled generators and which today serves as a test hall for civil engineers at the Technische Universität Berlin. What emerged in Germany in the form of modern and lightweight steel structures thus also found its way to distant Africa.

The remains of the steel landing stage in Swakopmund, the jetty, still stand today and support a restaurant at the very front over the water, where we enjoy fresh fish and the sunset over the Atlantic Fig. 9. "Colonial nostalgia is spreading. The exploitation of African workers and the actual purpose of the port constructions as German bridgeheads for the subjugation and colonisation of a country remain hidden from most visitors." [7]

German civil engineers continued to be involved in Namibia later on, and I would like to mention a connection to the Technische Universität Berlin. Karl Otto Ludwig Klaus Dierks (1936–2005) was a civil engineer who graduated from the Technische Universität Berlin in 1963 with a degree in civil engineering. He also received his doctorate from the university in 1992. He spent most of his life in Namibia and, after the country gained independence, made a name for himself in road construction and politics (he was a member of SWAPO). For over 20 years, he campaigned for the construction of the Katima-Mulilo Bridge connecting Zambia and Namibia and was able to witness its completion (see also Day 66). [8]

9

Fig. 9 The former landing stage, Swakopmund Jetty

10

11

Figs. 10–11 Namutoni Fort in Etosha National Park and a disturbing plaque at the entrance

Day 53, 6 May – Sesfontein

In the interior of the country, the German colonial forces built around 30 military installations, often rectangular fortifications with a walled courtyard and defensive towers at the corners. "The buildings were constructed from fieldstone and air-dried bricks" [9] and were thus markedly less advanced than the sophisticated structures of Namibia's railway and port facilities, which were built at the same time. The fortification in Sesfontein is now a hotel with a pool and roof terrace on one of the towers. We sit up there with a cold beer in our hands, framed by a Namibian and a German flag, and look through the battlements onto the sandy plain sparsely covered with acacia trees around the currently dry Otjitaimo River. Far from the beaten track and indeed far away from its initial use, we spent a very comfortable night here in Sesfontein. The concept of reuse has been implemented, and today the descendants of the former occupiers at least pay for their stay.

There is also a German fort that serves as an 'attraction' in the north of the country at Namutoni Rest Camp in Etosha National Park. I immediately indulge in fantasies of conversion and plan to repurpose the fort into a beautiful residential complex. But when I read the historical plaque at the entrance, which provides insight into its original use, I sober up and abandon the idea Figs. 10–11.

Day 58, 11 May – Ongwediva

Oshakati, Oshikango, Ondangwa – no, we have to go to Ongwediva, which is also in this area. Here, in the far north, a few kilometres from the Angolan border and 700 kilometres from Windhoek, you can study civil engineering at the José Eduardo dos Santos Campus of the University of Namibia (UNAM) in the recently opened 'German wing' of the new university building Figs. 12–13. The building was financed by Germany and for a time also supported with personnel – engineer Joachim Lengricht from Berlin was on site for many years. He put me in touch with young associate professor Philemon Arito. Philemon comes from Kenya, earned his doctorate in South Africa, and now teaches structural engineering at UNAM with great enthusiasm. He will visit us in Berlin for two months over the New Year period 2023/2024 to learn about bridge design at schlaich bergermann partner. It will be a very pleasant, though far too brief, collaboration. Philemon has also written a short text for this book, and hopefully we will manage to maintain regular contact.

Petrina Johannes, the young, energetic dean of the Faculty of Engineering, repeats what I hear time and again in Africa: "We need simple, decent, and affordable housing." For her, "sustainability must be geared to people's needs. Due to the hot climate, air conditioning is essential". When we visit her, the thermometer actually shows over 35° Celsius. UNAM has no research funding from the government and it is difficult to obtain other funds because, as Petrina Johannes confirms, the economy and the construction industry are not doing well. Currently civil engineering students at this small university are migrating to other fields such as electrical engineering. There is one full professor and three senior lecturers for the approximately 100 students in the bachelor's programme. Some teachers, such as Philemon Arito, come from other African countries and praise the fact that they have better career opportunities in Namibia and are hired on the basis of their qualifications. Elsewhere, the rule is often: "If you are from the wrong tribe, you don't get a job."

I had previously visited the much larger Namibia University of Science and Technology (NUST) in Windhoek, which emerged from the Polytec of Namibia in 2012. There, students graduated with a diploma that is awarded at technical colleges in Africa, which has a different status from the German university diploma that so many in Germany mourn the loss of. Every year, around 100 almost exclusively black students begin studying civil engineering in Windhoek. Those who visit NUST usually stay at the NUST Hotel School, built in 2003 by Kerry McNamara Architects from Windhoek. This pleasant mini-campus – consisting of individual building modules around small, shaded courtyards – for aspiring hoteliers and chefs is located right next to NUST. There I meet colleagues from the University of Wuppertal (BUW), who tell me about an interesting cooperation project with NUST. To avoid the high cost of having to document the road conditions of this sparsely populated country with measuring vehicles, sensor data from rental cars is to be evaluated in order to assess the quality of the road surfaces and respond more quickly to damage. [10]

12

13

Figs. 12–13 The UNAM campus in Ongwediva and Philemon Arito in the concrete laboratory there

Together with Aachen University of Applied Sciences, BUW also oversaw the transformation of Polytec into NUST. Since then, Aachen University of Applied Sciences has been offering a double degree in civil engineering with NUST. NUST also has close ties to South Africa, as most of its teaching staff were trained there. The main areas of teaching are hydraulic engineering and, due to the uranium mines in Namibia, geotechnical engineering.

1

Fig. 1 The salt road along the Skeleton Coast

Civil Engineering and Building Construction in Africa
Namibia Prof. Dr. Philemon Arito,
UNAM, Ongwediva, Namibia, January 2024

'Namibia's construction industry is still in its infancy. The training of civil engineers within Namibia's borders began in 2009 with the University of Namibia (UNAM) and was further advanced in 2015 with the transformation of the Namibia University of Science and Technology (NUST) from a technical college into a university. UNAM and NUST are the only universities in Namibia that train civil engineers who can be registered as licensed engineers with the Engineering Council of Namibia (ECN). In addition, there are five state-run vocational training centres (VTCs) and many private colleges that train technicians and skilled workers for the construction industry. UNAM and NUST produce a maximum of 150 graduates from their bachelor's degree programmes in construction each year. This number is well below the current national requirement of 1,000 licensed civil engineers.

This discrepancy has led to an acute shortage that jeopardises the realisation of the Namibia Vision 2030 programme and the implementation of National Development Plan 5 and the Harambee Prosperity Plan II. The Lüderitz municipal administration, for example, is one of many local authorities in Namibia that lacks qualified and experienced civil engineers to review and approve construction plans. Due to the shortage of experienced civil engineers, Namibia is, therefore, forced to rely heavily on foreign engineers, most of whom come from South Africa and other countries in the Southern African Development Community (SADC).

About 70 per cent of the civil engineering firms registered with the Association of Consulting Engineers of Namibia (ACEN) are small firms, meaning they employ fewer than ten licensed civil engineers. 75 per cent of these are Namibian-owned. 90 per cent of engineering firms employing more than 30 licensed civil engineers are international. At least 80 per cent of international engineering firms in Namibia are South African-owned. 90 per cent of large engineering firms employing more than 20 civil engineers are multidisciplinary. They work on projects across the entire spectrum of engineering: civil engineering, mechanical engineering, and electrical engineering. 80 per cent of construction companies in Namibia are owned by Namibians and are mainly involved in small to medium-sized projects such as low-rise buildings.

Large-scale projects such as the expansion of Hosea Kutako International Airport Road, the port expansion of Walvis Bay, and the construction of the Neckartal Dam are usually undertaken by foreign construction companies. 70 per cent of the foreign construction companies involved in megaprojects such as the expansion of Hosea Kutako International Airport Road come from China. Namibia's second-largest cement manufacturer, Cheetah Cement, is also Chinese-owned. Namibia does not have its own set of rules for building construction. Buildings and bridges are usually designed according to South African National Standards (SANS). However, depending on their affiliation, history, and the country in which the owner was trained, some companies use other sets of rules such as Eurocodes or British or American standards.'

DEMOCRATIC
REPUBLIC
OF THE CONGO
Boma
Matadi
M'banza Kongo
Uíge
Dundo
Luanda
Morro da Cruz
Kilamba
Quilundo
Cabo Ledo
Sumbe
ANGOLA
Luena
Lobito
Catumbela
Benguela
Huambo
ZAMBIA
Mapunda
Lubango
Bruco
Day 61, Cuito Cuanavale
Kawaya
Tômbwa
Day 63, Cuatir Nature Conservation Project
Helao Nafidi
Oshakati
Ongwediva
Olukonda
Kasane
Victoria Falls
BOTSWANA
Maun
NAMIBIA
Sowa
Twyfelfontein
Omaruru
Okahandja
Windhoek
Gobabis
0 km
250 km

Angola
12–17 May 2022

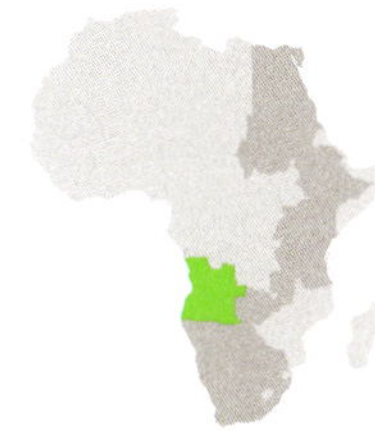

Angola is the only country for which we couldn't find a Lonely Planet guidebook, and in fact we don't see a single tourist during our five-day stay in the south of the country. Once it's clear that we're allowed to enter and we've spent five hours dealing with border formalities, we drive as far as possible through the sparsely populated southern provinces of Cunene and Cuando Cubango. A litre of petrol costs only 30 euro cents, but at the few petrol stations we find, there are long queues of moped riders who fill their small yellow canisters as soon as petrol becomes available. Everyone we asked in Namibia advised against travelling to Angola, but none of them had actually been there before. As is so often the case, fears and prejudices initially unsettle us, but we find a scenic and safe country with extremely friendly people. Minor adjustments are necessary: we get used to driving on the right again, try to understand Portuguese, and enjoy Portuguese wine with the funge (manioc porridge) served everywhere.

Day 61, 14 May – Cuito Cuanavale

The country is still marked by decades of proxy and independence wars between the MPLA (Movimento Popular de Libertação de Angola), supported by the Soviet Union and Cuba, and the UNITA (União Nacional para a Independência Total de Angola), with the USA and South Africa on its side. The national parks are empty because the wildlife populations served as a source of food during the long civil war. Large parts of the south are contaminated with over 15 million landmines. [1] There is neither a campsite nor a hotel in Cuito Cuanavale. We follow a sign saying 'camp' and arrive at the mine clearance camp run by the international Halo Trust foundation, where we find shelter for the night. A young English woman runs the camp. Her job is to drive heavy Russian trucks into the terrain to identify and defuse mines.

Between 1987 and 1988, the largest battle in Africa since El Alamein was fought on the river-laced, tall-grass plains around Cuito Cuanavale. The military arm of the MPLA, the FAPLA, supported by Cuban and Soviet forces, and UNITA, with South African soldiers at their side, engaged in a year of brutal slaughter in this godforsaken area. A monumental sculpture in North Korean style, inaugurated in 2017, serves as a visible reminder of these fierce battles, which claimed thousands of lives. [2, 3] This curious monster, in front of which two giant warriors cast in bronze stretch up to the sky, at least offers a viewing platform Fig. 1. From above, we see an open-air park where tanks, aeroplanes, and rocket launchers are proudly displayed as well as a museum and even a runway, which is surely only used for occasional commemorative ceremonies Fig. 2. Visitors there are none.

1

2

Figs. 1–2 The war memorial from North Korea in Cuito Cuanavale and a view from above of the huge car park where our car stands alone

3

4

Figs. 3–4 In the countryside: MPLA and UNITA villages with their respective flags

The battle had no clear winner, even though the adjoining museum claims victory for Angola and makes no mention of the more than 30,000 Cuban soldiers sent by Fidel Castro to support the MPLA.

Since the elections in 1992, the MPLA has officially ruled the country, although UNITA continued to operate underground until 2002. Today there seems to be a largely peaceful coexistence between the former civil war parties in Angola, albeit still separated by location. Wherever we go, there is always a flag flying in the village: either the red and black flag of the MPLA with the yellow star or the red and green flag of UNITA, which shows a rooster in front of a setting or rising sun Figs. 3–4.

Here in the south, we repeatedly come across a stretch of well-maintained road. We see new bridges that have been built right next to those that were destroyed in the war. It is clear that Angola's economy has been on the upswing since 2002. We also notice new wells and schools, which we identify as having been built by UNICEF. Why can't this (oil-rich) country do this on its own? This is certainly also due to the 38-year rule of President José Eduardo dos Santos (MPLA), who stepped down in 2017. He dies a few weeks after our stay in July 2022. Corruption was rampant during his tenure. [4] Hopefully 'Homo homini lupus' and kleptocracy are now a thing of the past in Angola, but there is still a lot of reconstruction work to be done Figs. 5–6.

In Kuvango, a special arch bridge, probably a result of economic euphoria, makes me pause. A skew arch bridge is not very practical, but it is structurally challenging. As I walk around the bridge, I realise that caution was exercised and a beam bridge with a non-load-bearing arch was chosen instead Figs. 7–8. In this city, I also photograph one of the multi-storey concrete buildings in Chinese style that we keep noticing here (→ Considerations – China).

5

6

7

8

Figs. 5–6 In the countryside: concrete bridge destroyed in the civil war with a new parallel steel truss bridge and the remains of a small concrete bridge in southern Angola

Figs. 7–8 A striking road bridge in Kuvango with a non-load-bearing arch

Road construction also seems to have been handed over to China. The EN280 along the Cubango River between Matala and Cuito Cuanavale is a brand-new asphalt road, but there is hardly any traffic on it. A Chinese road project from Menongue south to the border with Namibia at Katwitwi has obviously been abandoned, and the many machines intended for use on it are rusting away halfway along the route Fig. 9. It is a pitiful sight to see so many expensive machines falling into disrepair! It takes us almost two days to cover the 350 kilometres along this stretch of the EN140 south.

Day 63, 16 May – Cuatir Nature Conservation Project

Sustainable construction and eco-tourism? Stefan van Wyk deserves a mention here. The cosy Villa Menongue guesthouse in Menongue and his buildings in the private Cuatir Conservation Project further south are constructed from local materials and powered by solar energy. The Namibian-born pilot has settled in Angola and is attempting to improve the situation of the local population by promoting "development through tourism". [5] He even buys zebras in Namibia to 'repopulate' his private Cuatir Nature Park. He had probably hoped for a paved EN140 when he built here, deep in the south. Now, unfortunately, the Cuatir Park is quite difficult to

9

Fig. 9 Abandoned Chinese road construction machinery

reach. We take a small ferry across the Cubango. Further south in Botswana, it seeps into the Okavango Delta. We drive along a soft sandy track through a sparse fairy-tale forest with trees whose exotic names we later learn from Stefan: silver clusterleaf (Terminalia sericea), Zambezi teak (Baikiaea plurijuga), wild teak (Pterocarpus angolensis), and African rosewood (Guibourtia coleosperma). Arriving at the van Wyk forest camp, we just catch sight of the red sun setting over the hazy grass plain by the Cuatir River. Here you can go on beautiful hikes and canoe trips, provided you have a good mosquito repellent with you. As the only guests, we have our pick of the newly built, locally constructed wooden cabins with solar panels and wood-fired outdoor showers, and we also have the well-equipped communal kitchen to ourselves. Stefan builds the supporting structure of his houses with rubber tree trunks that have been treated against termites and imported from South Africa. Otherwise, he uses local teak. He conserves the rare African rosewood because, he complains, the Chinese would try to exterminate it in Angola, "chopping down each and every one" for export to China Figs. 10–13.

Once again, a truism proves itself. Where the roads are good, the situation of the population improves because people can transport their products more easily and goods

10

11

12

13

Figs. 10–13 The drive to the Cuatir Conservation Project, the main house with the communal kitchen, and impressions of the area around the Cuatir River

reach them more quickly. So, in addition to good roads, we repeatedly see local facilities where air-dried light-coloured mud bricks – sometimes mixed with cement – red-fired clay bricks or concrete blocks are produced and sold Fig. 15.

Away from the main routes are the huts of subsistence farmers, built locally from wood and clay. Here, root crops and cereals are usually grown. Massimo has worked for decades all over the world as an expert on food security in international cooperation for rural development. He explains to me that many of these subsistence farmers are dependent on aid because a household cannot cultivate more than two to three hectares of land by hand. Even with sufficient rainfall, the yield from these fields will not provide enough food to sustain the family throughout the year. There is no question of using the proceeds from sales to pay for education and health care. Most farmers do not have oxen to pull their ploughs, let alone access to improved seeds or machinery for cultivating their fields Figs. 16–17.

It is unbelievable. By 2025, 65 per cent of the world's unused arable land will be in Africa. The future of global food security, therefore, lies in Africa's hands, yet at the same time, many of the farmers here are not even able to feed

14

15

16

17

Figs. 14–17 A Portuguese snack bar and masonry work next to the road (14–15). Drying and threshing millet (16–17)

themselves. Too little is being spent on increasing their productivity, even though the technologies are available and have been successfully tested in Asian and South American countries. In 2022, Asia has the highest public spending on agriculture worldwide. It amounts to 77 per cent in nominal terms, while Africa's share is just 2 per cent. [6]

Massimo shakes his head: "This is the result of many years of costly development programmes that have failed to combat the factors of poverty. Unfortunately, we now have to conclude that most foreign aid initiatives are focused on humanitarian emergencies in order to feed a population that is increasingly suffering from food insecurity and living in unstable and conflict-ridden areas."

In Angola, around 80 per cent of farmers live from subsistence agriculture. [7] In Zambia, the next country on our journey, the majority of the rural population is in the same situation. [8] There, 60 per cent of farmers have to make do with less than two hectares of land each. [9]

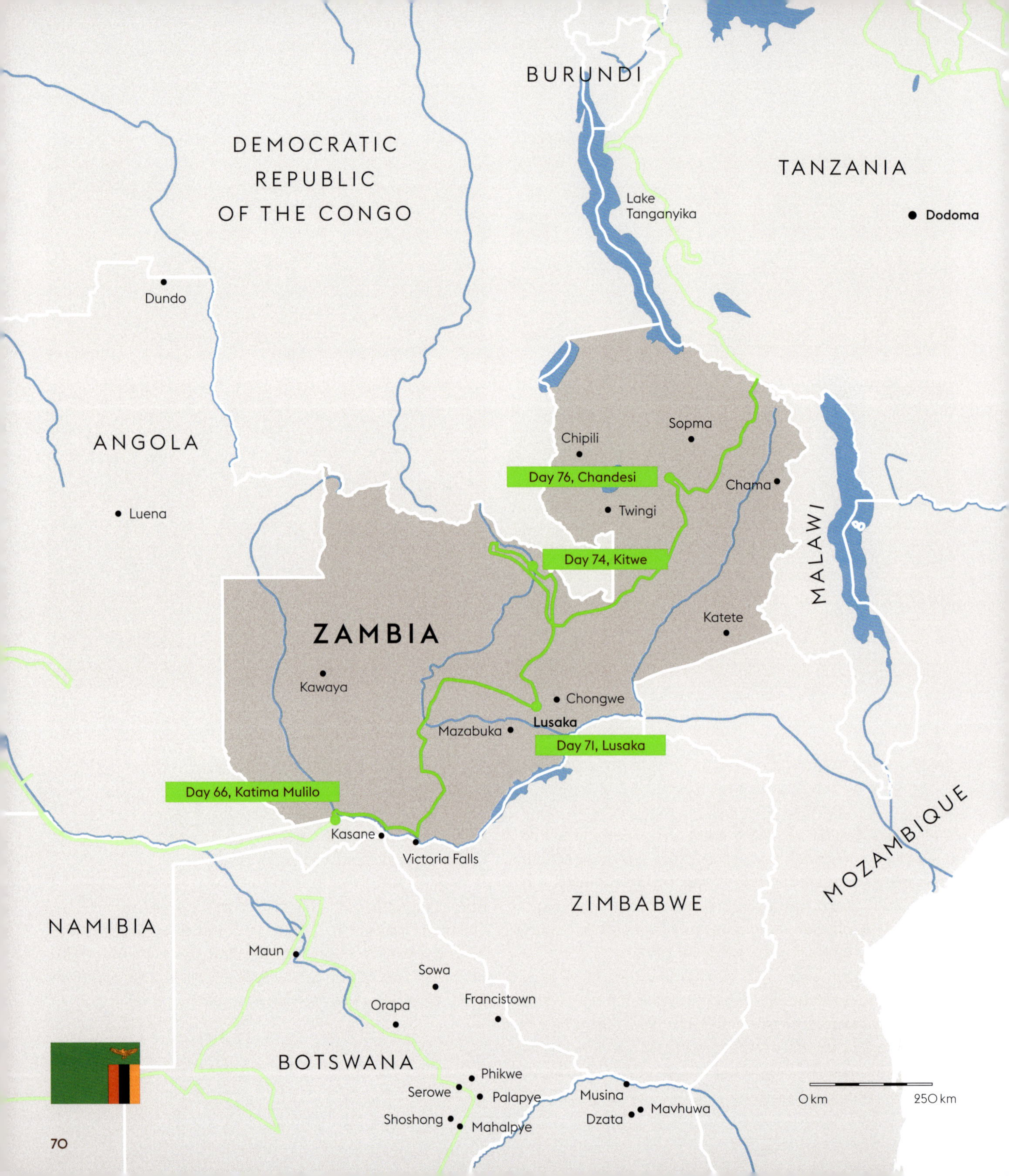
BURUNDI
DEMOCRATIC REPUBLIC OF THE CONGO
TANZANIA
Lake Tanganyika
Dodoma
Dundo
ANGOLA
Chipili
Sopma
Day 76, Chandesi
Chama
Twingi
Luena
Day 74, Kitwe
MALAWI
ZAMBIA
Katete
Kawaya
Chongwe
Lusaka
Mazabuka
Day 71, Lusaka
Day 66, Katima Mulilo
Kasane
Victoria Falls
MOZAMBIQUE
ZIMBABWE
NAMIBIA
Maun
Sowa
Orapa
Francistown
BOTSWANA
Phikwe
Serowe
Palapye
Musina
Mavhuwa
Dzata
Shoshong
Mahalpye
0 km
250 km

Zambia
19–31 May 2022

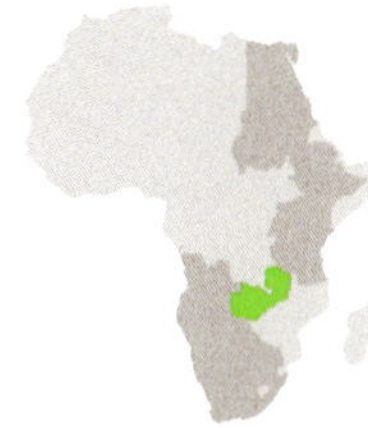

Zambia, known as Northern Rhodesia until gaining its independence in 1964, experienced a period of socio-economic development after the end of the one-party state (1972–1991). Nevertheless, it still remains on the UN list of least developed countries. However, it is rich in water, fertile land, and mineral resources. Several dams (Kafue 1 and 2 and Kariba) provide hydroelectric power and supply the country with electricity. In the north, the economic centre, copper is mined on a large scale.

Day 66, 19 May – Katima Mulilo

Colonial transport infrastructure, areas in south-eastern Angola that were mined after the civil war, and the railway connection from Zambia to the port of Dar es Salaam since the 1970s may explain why there is still no road connecting the south of the two countries. For us, this means a detour via Namibia's Caprivi Strip. After driving about 500 kilometres along the Cubango, we reach Zambia via the Katima Mulilo Bridge. It crosses the Zambezi, Africa's fourth-longest river, which we now follow. The prestressed concrete bridge, completed in 2004, was built by the construction companies Concor from South Africa and Hochtief from Germany. Klaus Dierks initiated the project for the 877-metre-long, 17-span concrete box girder bridge, and VKE (both of which we already know from Namibia) and the German engineering firm GAUFF designed it. The Nuremberg-based engineering firm GAUFF is very active in Africa and was also involved in the planning of Mozambique's Maputo-Katembe Bridge, Africa's longest suspension bridge.

However, the 'road' from here, north of the Zambezi, to Kazungula is so bad that it is hard to imagine any significant goods traffic to Zambia. We are familiar with routes like this. If asphalt roads are not maintained, potholes become so numerous and deep over time that you can only proceed at walking pace. At some point, I feel so sorry for the shaken car that I drive 'half and half': two wheels on the asphalt, which is in slightly better condition at the edge of the road, and two wheels on the unpaved and lower-lying hard shoulder. After a while, I can no longer bear the uneven driving and veer left onto the parallel dirt track that cars driven by similarly desperate people have formed over time. As soon as we can no longer stand the swirling dust, we return to the asphalt, and the cycle begins anew Fig. 1.

We should have taken the southern route through Botswana with another border crossing. Why is no one repairing the northern road? Why is traffic heading towards Zambia being forced south through another country? Do the concessionaires of the Kazungula Bridge have a say in this?

In May 2021, Zambia and Botswana inaugurated the 923-metre-long, seven-span Kazungula Bridge over the Zambezi with spans of up to 129 metres Figs. 2–3. It is probably the first multi-span extradosed bridge in Africa. In this variant of the cable-stayed bridge, short pylons and, therefore, very flat cables support a rigid concrete box girder, which forms the superstructure of the bridge. The cables mainly serve to support the dead weight of the superstructure. They hardly feel any traffic loads because the rigid superstructure absorbs the varying loads from cars and lorries.

1

2

3

Figs. 1–3 Condition of the road from Katima Mulilo to Kazungula and the Kazungula extradosed bridge over the Zambezi

As the cables, therefore, have fewer variable loads to bear, they can be used and anchored more economically. For spans of 100 to 200 metres, the rather elegant extradosed bridges are among the most economical bridge structures. The Kazungula Bridge from Zambia towards Namibia, Botswana, and Zimbabwe in the south became a political issue because neither Namibia nor Zimbabwe wanted to contribute to the costs. As a result, the bridge over the Zambezi River bypasses the countries unwilling to pay and ends up on the 135-metre-wide strip of land belonging to Botswana. [1] It was financed by the Japanese International Cooperation Agency and the African Development Bank. The South Korean company Daewoo was responsible for constructing the bridge with its concrete superstructure.

A few days later, naturally we also visit the grand old lady of the area: the Victoria Falls Bridge near Livingstone, built in 1905 Figs. 4–5. This 156.5-metre-wide steel arch bridge connects Zambia with Zimbabwe. At the proud age of over 100, it is still capable of carrying railway loads, and every tourist coming from Zambia who wants to visit Victoria Falls will cross it. On both sides of the bridge, you can admire the Zambezi River as it thunders down a 100-metre-deep and only 50-metre-wide basalt gorge, spraying mist hundreds of metres into the air. Incidentally, the bridge was inaugurated by Charles Darwin's son, Sir George Darwin, who was then president of the British Association (today the Royal Society). [2] During my later research, I learned that engineers Douglas Fox and Ralph Freeman were involved in the planning. The firm Freeman and Fox later made a name for itself with the design of important bridges such as the first Bosphorus Bridge in Turkey and the Humber Bridge in northern England. In the 1990s, I had the privilege of working on the Ting Kau cable-stayed bridge in Hong Kong with the last scion of this engineering dynasty, Ralph Freeman's grandson Tony.

4

5

Figs. 4–5 The Victoria Falls Bridge in 2022 and in 2014 with my father on his last long-distance trip

6

7

Figs. 6–7 The UNZA library in Lusaka with internal access via a ramp

Day 71, 24 May – Lusaka

In Lusaka, the capital of Zambia, it is important to choose your hotel carefully. We booked the cheaper of the two Marriott hotels, Protea Lusaka Cairo Road rather than Protea Lusaka, and ended up in a somewhat unsafe area. Every time I walk to the car park 50 metres away, where our car is safely parked, a hotel employee accompanies me at a discreet distance. Is it really that dangerous here, or does he want to supplement his tip? I decide not to pursue the question, enjoy my bodyguard, and reward him generously.

With around 30,000 students, the University of Zambia (UNZA) is the largest public university in the country. It opened in 1966, two years after the country gained independence, and the first civil engineering students graduated in 1972. The School of Engineering, which includes the Department of Civil Engineering, is located on the main campus on Great East Road.

I am greeted by African modernism. The university library, made of exposed concrete without any ornamentation, forms the heart of the campus, which was laid out in 1966 according to the plans of a team of architects led by Cape Town-trained architect Julian Arnold Elliott [3]. A ramp winds its way through the interior of the square building over three floors, leading to the various library and study rooms Figs. 6–7. Why was the exposed concrete painted green here?

My colleagues in the nearby engineering building, which was added later, have announced my arrival to their students, and I am allowed to give my lecture on bridge design in a well-equipped and well-filled lecture hall. I take the view that integral bridges, monolithic structures without bearings, and expansion joints are not only desirable for Germany, but may also be of interest to Africa. Not only are they elegant and resource-efficient because they are slimmer than single-span bridges, but they are also more robust, durable, and low-maintenance. The approximately 30 per cent of students who are female also participate in the subsequent discussion on sustainable construction. Throughout Africa, the proportion of female students in bachelor's degree programmes in civil engineering is as high – or low – as it is here in Germany. There is great interest in bridge construction, but apart from South Africa and Egypt, bridge construction is not yet taught on the

8

Fig. 8 Sale of concrete hollow blocks at the roadside

continent. Unfortunately, I am told that there is also little cooperation with industry and not much money available for research. The laboratory still contains equipment manufactured in the Soviet Union.

The housing issue is urgent, and research is needed in this area. The current deficit of 1.5 million housing units in Zambia will more than double by 2030 [4] (in the larger neighbouring country of Tanzania, the housing deficit has already reached three million units [5, 6]). Travelling through Zambia, I repeatedly notice small precast plants selling hollow concrete blocks right by the roadside Fig. 8. Compared with solid concrete blocks, these lightweight blocks promise material-saving and well-insulated buildings. There is abundant limestone in Zambia, which is why the cement industry is so strong. The internet lists at least six cement factories. In 2022, the price of cement in Zambia is US$7/50 kg bag (in Germany it is approx. €10/50 kg).

9

10

11

Figs. 9–11 Charcoal burners in a forest close to the border with Congo near Chingola, Zambia

Day 74, 27 May – Kitwe

On a trip north towards the border with Congo (DRC), we get lost in the forest and meet charcoal burners who show us the way back. By chance, we learn where and how the charcoal sold everywhere along the roadside is produced. The charcoal burners cut clearings and char the tree trunks on the spot in large piles under a protective layer of clay Figs. 9–11. Charcoal is mainly used in cities because it produces little smoke when burned. In rural areas, wood is still burned for cooking and washing. This deforestation is certainly not entirely legal, but we hear that there would be huge outcry if the charcoal burners were banned from doing their work. In Kitwe or Lusaka, bottled gas would probably not be more expensive and would certainly be more environmentally friendly than charcoal, but who would retrain the charcoal burners? And for what profession? Could the wood be used for construction instead of being burned? Alas, the poor forests of Zambia.

Founded in 1987, Copperbelt University in Kitwe is the second largest university in the country with around 10,000 students. Professor Senthil Kumaran, who came here from India, looks after the 30 civil engineering students and 25 technicians who start here every year. The tall, moustachioed man with dark glasses earned his doctorate in Chennai in southeast India and then taught in Rwanda. For the past two years, he has been a professor in the Department of Computer Engineering in Kitwe, the main copper mining centre in Zambia, just a few kilometres from the Congo (DRC). Because our engineering consultancy has already designed several bridges in India and because I remember Rwanda in 1983, we immediately strike up a conversation. He praises Rwanda as an example of a country where university members sit on government committees. This type of political consulting does not yet seem to exist in Zambia. Senthil Kumaran puts me in touch with colleagues in Rwanda and Burundi, and during the course of our journey, this amiable man repeatedly checks in with us via WhatsApp to see how we are doing. He wrote the text on construction in Zambia at the end of this chapter.

Senthil Kumaran confirms that concrete is also considered the material of choice for building construction in Zambia. He says that the slag produced during copper production can be mixed with cement to make blast furnace cement from the copper industry. However, he also points out the damage caused by copper mining, as copper processing releases large amounts of sulphur, which leads to acid rain and corrosion on buildings in the region. In addition, mine waste pollutes the groundwater Figs. 12–13. In his opinion, this is where research needs to start: climate change is noticeable, and the country is becoming increasingly dry. "Engineers are trained in the Western tradition for Western conditions. This is a tragedy, because local engineers should be qualified to plan rural roads and a good water supply as well as to work with local materials."

Senthil Kumaran tells me that he is working with colleagues from all over Africa on a RILEM book about the use of agro-based materials as cementitious additions in concrete and cement-based materials (RILEM: International Union of Laboratories and Experts in Construction Materials, Systems, and Structures). Copperbelt University is also conducting research into the use of local building materials. The colleagues here are working with laterite bricks made from solidified laterite soil. This red granulate has a high strength, which can be further increased by adding cement, lime, or other binding agents.

Other ideas I hear about revolve around banana-fibre-reinforced plastic, ferrocement using mine waste, and cement reinforced with sisal fibres and agricultural by-products and waste such as straw. A 1 MW photovoltaic system is also under construction on campus, a collaboration between Copperbelt University, Copperbelt Energy Corporation (CEC), and the German Society for International Cooperation (GIZ). However, the government has not yet launched a solar power programme because the country still has enough electricity from hydroelectric power plants.

12

13

Figs. 12–13 Copper bars on trucks in the Copper Belt and a road deformed by them

Day 76, 29 May – Chandesi

In northern Zambia, heavy downpours during the rainy season dictate bridge design, as the swelling rivers repeatedly erode the piers. When we visit the construction site of the historic Chambeshi Rail Bridge near Chandesi, it is the dry season and the sun is beating down. Halfway between Mpika and Kasama, the Chambeshi flows sluggishly past the man-high, already brownish riverbank grass, heading toward Lake Tanganyika. The Jiangsu Gold Civil Construction Company from China is currently working here with Hopkas Engineering from Lusaka to repair one of the steel truss girders, which fell victim to an improperly loaded freight train rather than the water. The bridge has been open again since late 2022, and the roads that had to bear the burden of freight traffic between Zambia and Tanzania for two years while it was rerouted are now under less strain Fig. 14 [7].

A few hundred meters further on, an old German cannon stands on a pedestal. The monument, erected by the National Monuments Commission, commemorates Lettow-Vorbeck, the only undefeated German general of the First World War. At the end of 1918, he entered British Rhodesia with his last soldiers on the long march from German East Africa to German South West Africa. He had just left Kasama on his way to the Chambeshi ferry: "a charming little spot. Like an oasis, the little village with its many clean European houses lays in the African wilderness". [8] On this journey, he learned from the British on 13 November 1918 that the war was over and he had to surrender. The era of the 'Huns' in Africa was finally over.

However, some British people have remained to this day. One hundred kilometres further on, we reach Shiwa House, which lies enchanted in a park (see also introduction Fig. 21). The English aristocrat Sir Stewart Gore-Brown (1883–1967) had a stately residence built here in 1914, far away from roads and railway lines and surrounded by 12,500 hectares of land. In contrast to apartheid in South Africa, he pursued a policy of 'collaborative leadership'. This seems to have been well received in the country. To this day, he is the only white person in the history of Zambia to have been given a state funeral. [9] We cheekily stroll through the large garden with its old trees and manicured lawn to the door and are nevertheless warmly received by his descendants, who occasionally welcome tourists here and show us around. Oil paintings and antlers on the walls, heavy dark wood furniture, and a large hunting dog on the carpet: time seems to have stood still here Figs. 16–17.

14

15

16

17

Figs. 14–17 In north-eastern Zambia: the Chambeshi Railway Bridge, the Lettow-Vorbeck Memorial, a plaque at Shiwa House, and a view of the Shiwa House dining room

18

Fig. 18 Breakfast at the Africanza Hotel near Copperbelt University

Construction and Building Engineering in Zambia, Africa
Prof. Dr. G. Senthil Kumaran,
Copperbelt University, Kitwe,
Zambia, February 2024

'Construction in Zambia encompasses a range of activities, from the planning and construction of residential buildings to large infrastructure projects. Zambia has regulations and standards governing the construction of buildings. For example, the National Building Regulation Act contains guidelines and standards for the construction industry that ensure the safety, durability, and sustainability of buildings. Zambia has invested in infrastructure development to promote economic growth and improve the standard of living of its population. This includes roads, bridges, dams, airports, and other important structures. All engineers in Zambia are licensed by the Engineering Institution of Zambia before they can practise professionally.

The University of Zambia (UNZA), Copperbelt University (CBU), and Mulungushi University (MU) are the only public universities in Zambia that train civil engineers. The number of graduates from these three educational institutions makes a significant contribution to the realisation of the Zambia Vision 2030 programme, the 8th National Development Plan (8NDP), and the African Union (AU) Agenda 2063. All local authorities in Zambia, city and municipal administrations, have qualified and experienced civil engineers who can review and approve construction plans.

Professional associations such as the Engineering Institution of Zambia (EIZ) play an important role in promoting excellence and expertise in the field of construction and civil engineering. They offer training, accreditation, and networking opportunities for engineers working in Zambia. The Zambian government has launched various initiatives to promote the construction industry, including infrastructure development programmes, training opportunities for local engineers, and partnerships with international organisations and donors.

Major infrastructure projects have been carried out by international construction companies with the help of grants, subsidies, and long-term loans. Many of the large engineering firms in Zambia are international. Buildings and bridges are usually designed to either South African National Standards (SANS) or British Standards (BS). Notable construction and building projects in Zambia include the construction of Kenneth Kaunda International Airport, the Kariba Dam, the Lusaka-Ndola Dual Carriageway, and numerous road and bridge projects throughout the country.

As in many developing countries, Zambia is undergoing rapid urbanisation, leading to an increased demand for housing and infrastructure in cities and towns. Civil engineers play a crucial role in the planning and construction of buildings to accommodate the growing population. Civil engineers in Zambia use a variety of materials and construction techniques that are adapted to local conditions. These include reinforced concrete, steel, masonry, and timber. Sustainable construction practices are also becoming increasingly important, with a focus on energy efficiency and environmental protection. Despite all the progress made, Zambia faces challenges in the building and civil engineering sector. These include inadequate infrastructure in rural regions, limited access to finance for construction projects, a shortage of skilled workers, and the need for the ongoing optimisation of standards and regulations.

Overall, construction in Zambia is a dynamic, continuously evolving sector that is crucial to the country's economic growth. The key to overcoming all challenges and unlocking its full potential lies in successful cooperation between the government, private sector players, and construction professionals.'

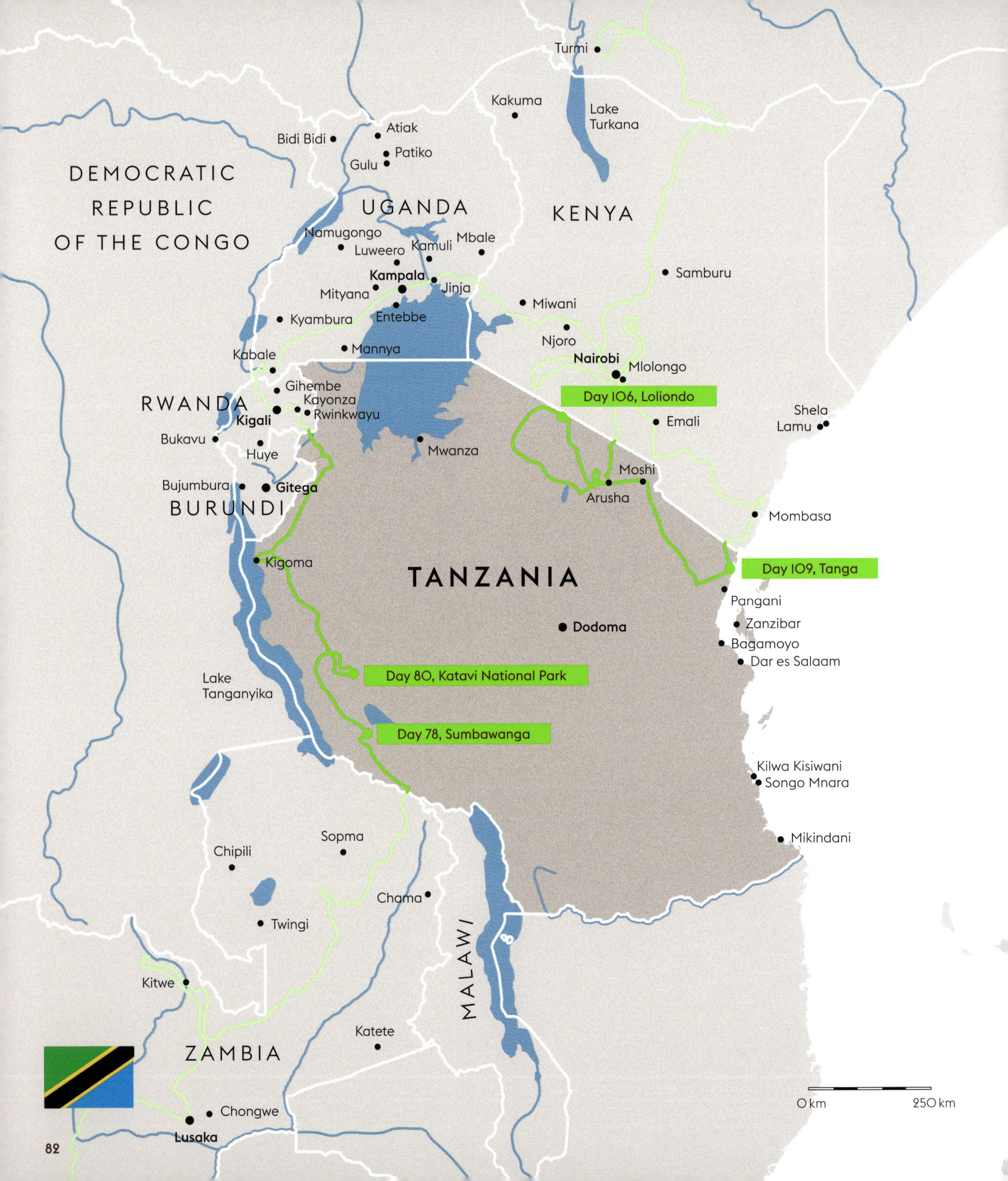

DEMOCRATIC REPUBLIC OF THE CONGO
UGANDA
KENYA
RWANDA
BURUNDI
TANZANIA
ZAMBIA
MALAWI
Turmi
Kakuma
Lake Turkana
Atiak
Bidi Bidi
Patiko
Gulu
Namugongo
Luweero
Kamuli
Mbale
Kampala
Jinja
Mityana
Entebbe
Kyambura
Mannya
Kabale
Samburu
Miwani
Njoro
Nairobi
Mlolongo
Day 106, Loliondo
Gihembe
Kayonza
Rwinkwayu
Kigali
Bukavu
Huye
Mwanza
Emali
Shela
Lamu
Moshi
Arusha
Bujumbura
Gitega
Mombasa
Kigoma
Day 109, Tanga
Pangani
Zanzibar
Bagamoyo
Dar es Salaam
Dodoma
Day 80, Katavi National Park
Lake Tanganyika
Day 78, Sumbawanga
Kilwa Kisiwani
Songo Mnara
Mikindani
Sopma
Chipili
Chama
Twingi
Kitwe
Katete
Chongwe
Lusaka
0 km
250 km

Tanzania
31 May–6 June: the west from Tunduma to Kigoma and 24 June–1 July 2022: the north from the Serengeti to Tanga on the Indian Ocean

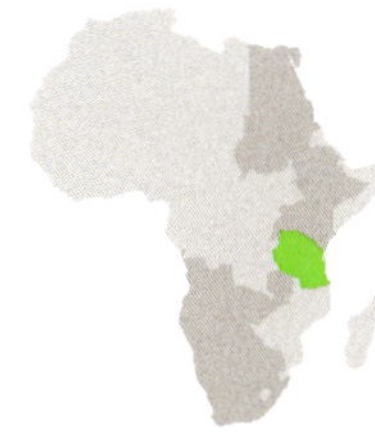

In Tanzania, we first visit the remote west on our way from Zambia and later the north from Kenya with its famous national parks and the Indian Ocean coast. I would have liked to see the capital Dar es Salaam again, but there is not enough time. My stays in the west and north alone show me how much the country has changed since 1983 when I travelled through it during the penultimate year of the rule of Julius Nyerere, the "Great Hero of Social Justice" [1], while it was still non-aligned.

Day 78, 31 May – Sumbawanga

Flashback: I met Massimo in Kigoma, Tanzania, 39 years ago in September 1983. After more than six weeks of travelling alone through Zaire, I had boarded the Swanza in Burundi's then capital of Bujumbura with my rucksack, heading for Tanzania. After a night-time journey across Lake Tanganyika, the second largest freshwater lake in the world after Lake Baikal, illuminated only by reddish dots along the shore of Zaire (probably bush fires), we arrived in Kigoma in the morning backlight. At the time I wrote: "Kigoma, nestled between a few hills on the shore, is not large: the harbour, the railway station, the terminus of the Dar es Salaam line, and a main street lined with small lodgings (hotels), and shops leading up the mountain to Mwanga. That's all." Two Italian couples, Massimo and Sandra together with Alberto and Francesca, had just arrived by train. They were students, just as dishevelled and dusty as I was, the same age, and also backpacking. Together we looked for accommodation. After that, we stayed together for six weeks, exploring Dar es Salaam and Zanzibar and partying quite a lot.

I felt much safer in the group, and because we could pool our resources, renting a Land Rover – a previously unattainable luxury – became a possibility. First, however, we set sail again continuing south, unfortunately not on the Liemba. I read in my diary: "We are invited for a glass of wine by a first-class guest, the captain, and the chief engineer. He comes across as a super-smart mechanical engineer, with a Cambridge degree and letters of recommendation, which he shows us. But then something must have happened to him. Now he drinks and can only be a chief engineer." The next morning, we were taken ashore in a swaying, fully loaded dinghy to Kabwe in Tanzania, where half the village was already waiting for goods and marvelling at the rare tourists. "Large granite boulders form the shore, interspersed with small sandy coves, and where the wind and surf allow, reeds grow down into the water. Above it all floats the grey-blue African sky. At first, all the village youth are after us ..." Figs. 1–3.

From Kabwe, we travelled for eight hours in a packed bus, plagued by tsetse flies, through a very dry steppe landscape to Sumbawanga. At least there were two stops where we could stock up on peanuts, sweet bananas, and tea and wake up our numb bottoms. At the time, Sumbawanga seemed like a small Wild West village to us. The bars had plenty of Konyagi, the Tanzanian schnapps, but due to a lack of fuel, there were no connecting buses. Bread, sugar, and cigarettes were only available on the black market. The

1

2

3

Figs. 1–3 1983: Arrival on the shores of Lake Tanganyika (1 and 3) and cooking in a village on Lake Rukwa

then socialist country lacked foreign currency. At the bank, you could get 12 shillings for one US dollar, which made life unaffordable for backpackers. On the street, however, you could get between 50 and 70 shillings for your dollar. The rule was: only exchange money with Indians, Arabs, or Somalis, as everyone else could be a policeman. At night, you could meet with a bank employee who issued official exchange receipts for a 'fee'. In the event of one of the frequent police checks, the now clean shillings could no longer be confiscated. At the same time, we noticed that not only the Konyagi, but also the glasses from which it was drunk, as well as crockery, cutlery, and many other items were produced locally, i.e. 'made in Tanzania'. After a week 'stuck in Sumbawanga', we found a truck and endured an eight-hour journey on its flatbed along a dusty road, alongside a heavy spare wheel that bounced with every pothole and threatened to kill us, until we finally arrived in Tunduma.

Today, on 31 May 2022, we – two greying 62-year-olds – drive our own car on a new asphalt road with kerbs safely separating the footpaths and speed bumps before every village, from Tunduma to Sumbawanga in just three hours. This place from our memory has now become a pretty little town, where a petrol station with a membrane roof and plenty of petrol welcomes us. We purchase a SIM card at a telephone kiosk and immediately have internet access.

4

5

6

7

Figs. 4–7 The road to Sumbawanga (4), roughly the same spot in Sumbawanga in 1983 and 2022 (5–6), membrane roofs of the local petrol station (7)

The hotel, bank, and bars from back then have disappeared, and we choose one of the new hostels. At the bank, we now get 2,300 shillings for one dollar; there is no longer a black market Figs. 4–7.

The 223-kilometre road from Tunduma cost around 100 million US dollars. The US government co-financed it through the Millennium Challenge Corporation (MCC). The MCC website discusses the economic returns of projects like this in detail. [2] It attempts to calculate how the economic conditions of individuals, the population, and the economy in the area along the route have changed. I take my time looking at the tables, but I can't find any final assessments. Is it even possible to make assessments like this? In any case, there are many new detached houses with corrugated iron roofs along the new road. In the villages, small piles of stones line the roadside – concrete aggregates for sale – and nearby we see local brick factories. Bricks moulded from clay are stacked in rectangular piles, leaving pizza oven-shaped cavities at the bottom. Wood is then burned there until durable masonry is fired Fig. 8. The good road allows these products to be transported economically to nearby construction sites. And Sumbawanga? I am sure it has benefited.

The Tanzania Roads Agency (Tanroads) is responsible for the country's nearly 13,000 kilometres of trunk roads and 22,000 kilometres of regional roads. [3] While around two-thirds of Tanzania's trunk roads have now been asphalted, virtually all secondary roads are still unpaved.

North of Sumbawanga, we are back on unpaved roads. "We had to wait for a bus to be extracted from the mud. For the adventurous," warns a traveller on the iOverlander app about the main road (trunk road) towards Kigoma. Fortunately, this only applies during the rainy season, but by the time we arrive in Kibondo 650 kilometres further on, driving at walking pace in places, the dust from the dry track has settled into every nook and cranny of the car. But we have seen construction sites. The African Development Bank and the Tanzanian government are having the route upgraded by Chinese construction companies. Two colleagues from Tanroads, whom I later meet at the IABSE Congress in New Delhi in September 2023, tell me that 80 per cent of the route has now been asphalted Fig. 9.

One of them is Katelula Kaswaga, a young civil engineer from Tanzania who is very passionate about bridge construction, which he says is "close to my heart". At the congress, he enthusiastically reports on the three-kilometre-long Magufuli Bridge near Mwanza on Lake Victoria. The multi-span cable-stayed bridge is currently under construction, and its cables are to be illuminated at night in the national colours of Tanzania. Katelula is proud to be involved in the construction and that his country can offer three more cable-stayed bridges: the Nyerere Bridge, the Tanzanite Bridge, and the Kirumi Bridge, which was built in 1980. He plans to write a book about Tanzania's major bridges and provides me with a short report on 'his' bridge and bridge construction in Tanzania.

Day 80, 2 June – Katavi National Park

We experience adventure in the relatively small Katavi National Park, which covers 4,400 square kilometres. As it is off the beaten track and, in any case, there are hardly any tourists due to the Covid-19 pandemic and the recent rainy season, we have the park practically to ourselves. The drive to Paradise Island becomes a psychedelic journey through a zoological-botanical garden. Crocodiles doze with their mouths open on the banks of the Ikuu River, elephants and giraffes stride majestically through endless grasslands, and hippos, which lie in mud holes for hours, suddenly turn around their longitudinal axis when the mud dries and itches, so that they are once again covered in cooling brown mud. I would love to shout my happiness out of the window, but it remains closed because of the tsetse flies. The paths have practically disappeared, probably because no one has been here for a long time. We have to negotiate metre-deep ditches and lose part of our bumper when we hit the ground. We have to squeeze through thickets, where a rear-view mirror gets lost. Then we turn back. The car, and more so we ourselves, have reached our limits, and we really don't want to get stuck in this solitude Figs. 10–13.

8

9

10

11

12

13

Figs. 8–9 Brickworks near Sumbawanga and the dusty road from there to the north

Figs. 10–13 Katavi National Park: crocodile warnings should be taken seriously ...

Day 106, 28 June – Loliondo

The Maasai are probably the best-known ethnic group in Africa. Anyone who books a safari through the Serengeti will see these proud and strikingly dressed warriors and cattle breeders. There are currently less than two million Maasai living in East Africa. But their population is growing, and this is where the problems begin, as they and their cattle are moving ever closer to nature reserves and the tourism industry. The government, therefore, wants to resettle the Maasai yet again. When we are in the area, there have just been clashes with the police. Which is more important? The world cultural and natural heritage site, the spectacular Ngorongoro Crater with its unique wildlife? Or the 60,000 Maasai herders with their 350,000 cattle, who have always lived in this region? The question seems easier to answer for the area further north around Loliondo. There, too, we learn there have been recent clashes because businessmen from the United Arab Emirates have reserved large hunting grounds, which they see disturbed by the cattle of the Maasai living there. [4] There is a paved road from Loliondo Airstrip to the game reserve, but after that we struggle southwards in our car through a bone-dry lunar landscape on roads covered with centimetres of fine dust, which envelops us in an opaque cloud when there is a tailwind. I wouldn't want to swap places with the Maasai, who have to drive their cattle and goats through these areas unprotected and without air conditioning.

We meet Samuel Meiyan, a Maasai who works as a shop assistant in a hotel in Arusha for a monthly salary of 100 euros. As a city dweller, he does not wear a red and black or red and blue Maasai cloth around his waist, but arrives on his motorbike wearing jeans and a T-shirt. His stature also does not correspond to the common image of the tall, slender and willowy Maasai, who are known for walking long distances. Samuel is committed to ensuring that girls in his village near Monduli also receive an education like him and learn to read and write. Of course, this is not a given in the polygamous male society of the Maasai, and his eyes light up when he talks about his efforts to convince others. Samuel is certain that only education can give the Maasai the opportunity to successfully represent their interests. To finance these activities, he also brings tourists here, and we spend the night on a meadow covered with dry yellow grass near a Maasai farmstead, a boma, with a view of distant Kilimanjaro. Goat is on the menu for dinner. When Samuel and his friends bring in a live animal, I realise that I am about to witness my first live slaughter. Every part of the animal is used, and the kidneys are eaten raw immediately. We decline this, as well as the warm blood that is offered to us out of courtesy, but which they much prefer to drink themselves. We eat a little of the grilled meat and are glad that the rest will be taken to the village and distributed.

The next morning, I am allowed to see the boma. The small circular houses of the women and their children, the slightly larger house of the man, and the herd of goats are protected from attackers by a fence made of thorny branches. The wooden frames of the walls of the huts, which the women build themselves, are plastered with a mixture of earth, cow dung, and urine and have a thatched roof supported by radially and tangentially arranged branches. Inside is a mattress as a bed and a stove with firewood for cooking, although I don't see any ventilation. The sooty roof is kept free of insects by the smoke, but the carbon monoxide levels inside must be high. I read on the internet that the average life expectancy of the Maasai is 45 years Figs. 14–18.

Day 109, 1 July – Tanga

We drive along endless sisal fields at the foot of Kilimanjaro – my wife Paloma is accompanying me on this two-week leg of the journey – towards the Indian Ocean, constantly jostling with lorries whose superstructures bear inscriptions such as 'pray until something happens' or simply 'judge me'. We want to visit the city of Tanga, if only because of its melodious name, but above all because of its historic buildings.

This cosy, sleepy harbour town is home to a collection of East Africa's architectural styles. Merchants of the Sultan of Zanzibar, who took over Tanga in the mid-nineteenth century, the German colonial rulers, who built the Usambara Railway from here, and the British, who made money from the sisal trade, have left their mark on the area around Independence Avenue. There I also find Africa's most beautiful Toyota repair shop, and although there is nothing wrong with the car, I have it serviced so that I can take my time viewing the 1950s expressionist building Figs. 19–21.

14

15

16

17

18

Figs. 14–18 A boma, a Maasai farmstead, near Monduli, Tanzania, with two of its inhabitants

19

20

21

22

Figs. 19–22 Toyota repair shop (19–21) and Mkonge Hotel in Tanga (22)

We spend our last night in Tanzania at the Mkonge Hotel, a former German administration building from the 1890s with walls made of coral stone, as I read Fig. 22. The old house by the sea has witnessed the country's history; a brass plaque on the front door calls it the "Sir William Lead Memorial Building". Our room is in a modern extension, probably dating from the British era. High-quality dark hardwood doors and frames and a large bar, cooled by a slowly rotating ceiling fan, dominate the interior. I don't see any coral stones anywhere because the walls are covered with white plaster on the outside and ochre and brick-red plaster on the inside. Someone has made the effort, rather unsuccessfully, to colour the concrete blocks that line the path to the large garden and the sea in exactly these shades. We order a gin and tonic under the trees and let our thoughts wander as we listen to the clatter of the shrouds of the sailing boats in the nearby Tanga Yacht Club.

Planning and Construction of Bridges in Tanzania
MSc Katelula Kaswaga,
Project Engineer, J. P. Magufuli Bridge,
Tanzania, May 2024

'In recent decades, the Tanzanian construction industry has grown steadily and rapidly. Many long span bridges have been built. As super engineering projects, long span bridges must be built with the best possible technology in the shortest possible time. At the same time, they must be designed to have the longest possible service life and be as easy to maintain as possible.

Among the numerous long-span bridges in Tanzania, there are three cable-stayed bridges namely Kirumi Bridge, Nyerere Bridge, and Tanzanite Bridge. A fourth, the J. P. Magufuli Bridge in Mwanza, is currently under construction. With a total length of 3,000 metres, it will be the longest bridge in the country and the sixth longest bridge in Africa. It consists of a total of 66 spans. The multi-span main bridge is 'extra-dosed' with a span arrangement of 100 + 2 × 160 + 100 = 520 metres. The approach bridges are prestressed concrete beam bridges with a total of 62 spans of 40 metres in length. Their total length is 2.48 kilometres, with 1.24 kilometres (31 spans) on the Kigongo side and 1.24 kilometres (31 spans) on the Busisi side. The concrete box girder deck of the main bridge is 28.45 metres wide and provides space for four lanes with a width of 3.5 metres for vehicle traffic, 2.5-metre + 2.5-metre shoulders on each side for emergency parking and break-down assistance, 2.5-metre + 2.5-metre footpaths on each side and a 2.45-metre-wide central reservation. Also noteworthy are the three main pylons, which look like outstretched arms or open gates. They are intended to symbolise triumph and prosperity. The bridge was designed by Yooshin Engineering Corporation and Cheil Engineering Co. Ltd, both from South Korea, in collaboration with Afrisa Consulting Ltd and Apex Engineering Ltd, both from Tanzania. The construction is a joint venture between China Civil Engineering Corporation (CCECC) and China Railway 15 Bureau Group Corporation (CR15G). My role in this project is that of a bridge project engineer.

There are several engineering firms in Tanzania with experience in bridge design, such as Afrisa Consulting Ltd, UNITEC Civil Consultants, Inter-Consult Ltd, etc. Large bridges are usually built as part of joint ventures with international construction companies, mainly from South Korea, Denmark, Japan, Egypt, and Denmark.

Bridge engineering is only taught as a course or module in civil engineering degree programmes; currently no university in Tanzania offers a standalone degree in bridge engineering. I believe that a very limited number of students graduate with specialised knowledge in bridge design each year. This leads many Tanzanians to seek study abroad opportunities in countries such as China, Russia, Algeria, the United States, South Africa, Egypt, Germany, Canada, Israel, South Korea, and the United Kingdom, often supported by full or partial scholarships.

Typical bridge designs constructed in Tanzania include prestressed concrete bridges, reinforced concrete bridges, slab bridges, box girder bridges, composite bridges, extradosed bridges, and cable-stayed bridges. The biggest challenge is selecting the best materials for bridge construction and components, considering the availability of financial resources, durability, strength, sustainability, and the ongoing maintenance costs.

In my opinion, the construction industry, particularly bridge engineering, is growing rapidly, driven by the construction of many long span bridges. Furthermore, collaboration and partnership between local and foreign companies and engineers helps to increase the expertise of bridge engineers. Recently many young engineers and students have become interested in bridge engineering design and constructions. More than 20 engineers have also joined the International Association for Bridge and Structural Engineering (IABSE).

Bridge construction is not just my profession, but my passion. I hope my love for this branch of civil engineering will make me a true professional and inspire many engineers across the globe, particularly in my country, Tanzania, and across Africa.'

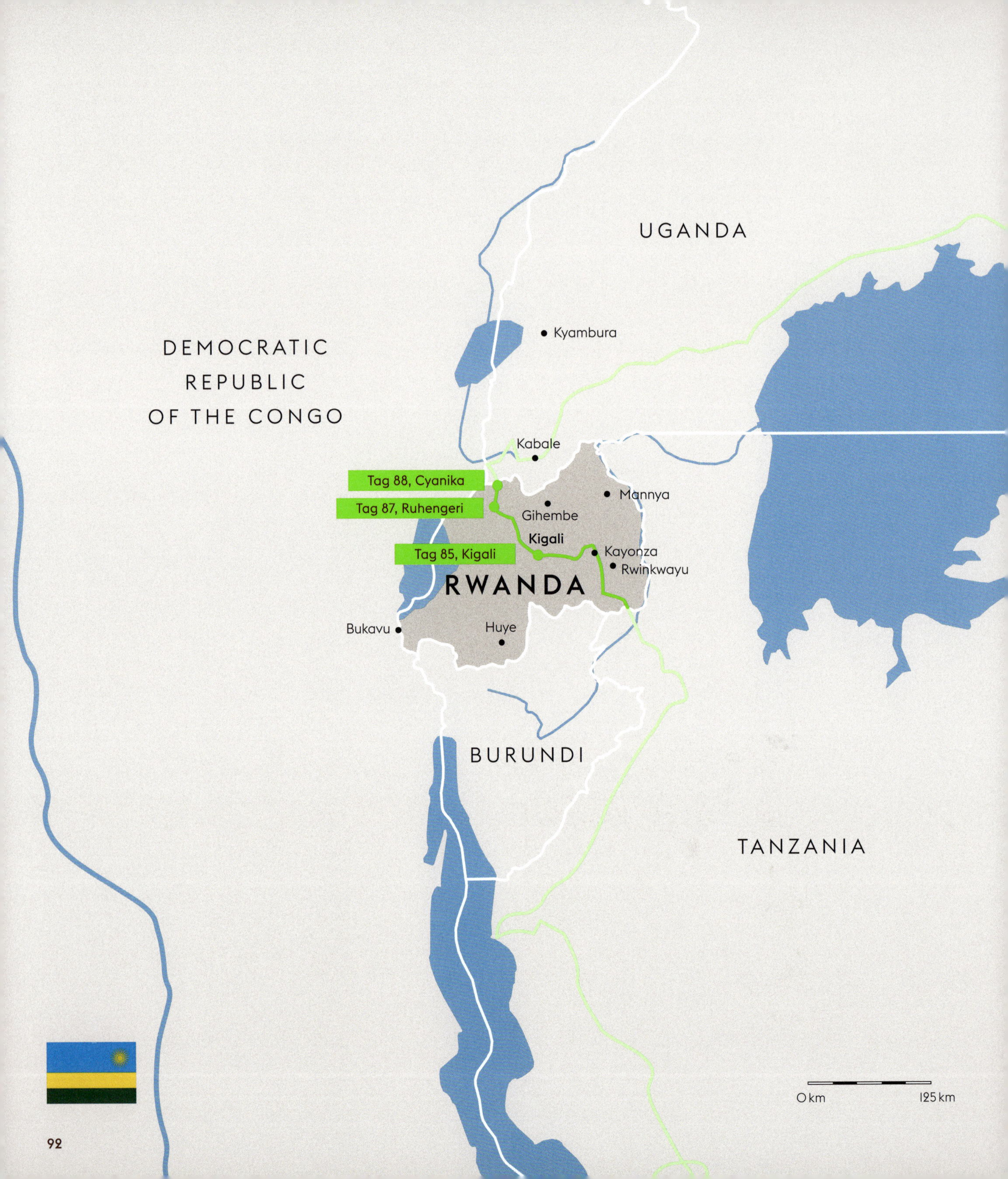
UGANDA
Kyambura
DEMOCRATIC
REPUBLIC
OF THE CONGO
Kabale
Tag 88, Cyanika
Tag 87, Ruhengeri
Mannya
Gihembe
Kigali
Tag 85, Kigali
Kayonza
Rwinkwayu
RWANDA
Bukavu
Huye
BURUNDI
TANZANIA
0 km
125 km

Rwanda
6–10 June 2022

The land of a thousand hills, green and fertile Rwanda is spotlessly clean. No plastic bags litter the roads alongside tea plantations and rice fields; they have been banned since 2008. From the border to the capital Kigali, the roads are swept, the drains are cleaned, and the grass is cut. It reminds me of a cleaning tradition in my home country, the Swabian 'Kehrwoche'. Pedestrians and cyclists often have paths that are neatly separated from road traffic. In addition, an old acquaintance – the graphite-grey pillar that flashes at us in Germany when we drive too fast – has made itself at home here, replacing the annoying speed bumps ('dead policemen') Figs. 1–2**.**

How terribly different it must have looked here 28 years ago! The 1994 genocide claimed the lives of around 800,000 people, mainly from the Tutsi minority. Since then, however, Rwanda's economy has been on the rise. Professor Umaru Wali from the University of Rwanda explains the reason for this positive development to me: "The genocide triggered a change in consciousness, and everyone is working hard to make up for lost time and catch up." He adds that the government is very good: it promotes performance and creates security. International observers, however, describe Rwanda less euphorically as a "successful developmental autocracy". [1]

1

2

Figs. 1–2 Attractive footpaths and cycle paths as well as speed cameras in Rwanda

With around 500 inhabitants per square kilometre, Rwanda, which covers an area of only 26,000 square kilometres, is more than twice as densely populated as Germany. Every tiny piece of land – no matter how steep the slope – seems to be cultivated here, and I wonder where the refugees that the UK is currently trying to deport to Rwanda are to be accommodated. How can this landlocked country, which is characterised by subsistence agriculture and remains on the UN list of least developed countries, finance its good infrastructure that is particularly expensive due to its topography? Political stability – and perhaps a guilty conscience for failing in 1994 – has certainly attracted a lot of loans and aid from around the world, as can be seen from the numerous NGO offices in the country. At least a few years ago, the sale of expensive ores such as coltan, smuggled into Rwanda from the conflict-torn eastern Congo, also contributed to the upswing. [2]

Day 85, 7 June – Kigali

Since March 2016, traffic travelling from Tanzania to Kigali, the capital of landlocked Rwanda, has had one less bottleneck to pass through. The route along the NR4 is a key logistics link [3] for supplying Rwanda from the port of Dar es Salaam on the Indian Ocean. Since that year, the One-Stop Border Post and the Rusumo International Bridge have opened, and we too benefit from the fact that traffic no longer congests at this border crossing. With a span of 80 metres and a superstructure height of only around three metres, the single-span girder bridge has become quite slender. Four steel girders, which act compositely with a concrete slab, form the two-lane, 13.5-metre-wide superstructure. The Japanese construction company Daiho built the bridge, which was financed by the Japan International Cooperation Agency (JICA). I want to learn more about civil engineering in Rwanda at the university in Kigali. It's only a four-hour drive away.

On the way to the University of Rwanda, I visit the Kigali Genocide Memorial, a harrowing must-see for any visitor to the capital. Located three kilometres from the city centre on steeply sloping terrain, the site features an amphitheatre next to the exhibition building, designed by John McAslan Architects from London. Unfortunately, the engineers did not provide it with an appropriate supporting structure for the roof, which consists of a heavy steel truss and a non-prestressed membrane Fig. 3. In the park, the names of many victims are immortalised on a black memorial wall that reminds me of the Vietnam Veterans Memorial in Washington. Heavy, inconspicuous grey concrete slabs cover mass graves in which around 250,000 murdered Tutsis are buried. Those who venture into the exhibition are confronted with bloody murder weapons, smashed skulls, explicit photographic documents, and comparisons with similar places in Germany, Japan, Cambodia, and Bosnia. Indescribable cruelty, indescribable suffering – I cry.

The state-run University of Rwanda, with 30,000 students, formed in 2013 from the merger of various universities and is considered Rwanda's largest educational institution. There, dressed in a black suit and white shirt, I am greeted by Professor Umaru Wali, Dean of the School of Engineering, College of Science and Technology at the University of Rwanda. He wrote his doctoral thesis in Moscow in the field of hydraulic engineering, and he helps me navigate the jungle of institutions here. The School of Engineering has three departments, one of which is the Department of Civil, Environmental, and Geomatic Engineering. He takes me there to meet Dr. Abaho Gerushome, the Head of Department (HOD), and his colleagues. We meet in a large windowless meeting room equipped with a projector and screen at the East African Institute for Fundamental Research (EAIFR). Sitting around a heavy oval wooden table, we introduce each other to our research and teaching.

Civil engineering, which has been taught here since 1960, accepts up to 100 students annually (the proportion of women, currently 15 per cent, is to be increased to the Rwandan standard of 30 per cent). The department initially offers a year of training, during which students can attend technical courses such as welding classes, followed by a four-year bachelor's degree and finally a two-year master's degree. After the third year, the bachelor's degree includes a ten-week internship in industry. I hear that money is tight and there are no research laboratories yet. Teaching focuses on the basics, building construction based on British standards. "Housing is the main problem", I hear again and again here, too. I also learn from colleagues that building with mud bricks is prohibited in the capital with its high-rise buildings; concrete is used instead, as the city is to be modernised. In the countryside, I often see single-storey reinforced concrete frames filled with masonry. The painting often seems to be done by sponsors Figs. 4–5.

I would like to learn more about local bridge construction. The mountainous country with so many rivers and valleys obviously needs it, and the moderate climate without large temperature fluctuations and without snowfall should allow for economical and material-saving bridge construction. However, I cannot get much help because bridge construction, which is still an expensive luxury, is not yet taught. The supreme discipline of civil engineering still seems to be left to foreigners here. At least I learn that for bridge construction only grade C50/60 concrete can be achieved because the compressive strength of aggregates in Rwanda is not very high. There are also complaints that there are no funds available in the country for maintenance.

Day 87, 9 June – Ruhengeri

The Institut d'Enseignement Supérieur de Ruhengeri, or INES Ruhengeri for short, is evocative of the former official language, French. Since 1994, there seems to have been a shift away from France, with English becoming the dominant language. [4] This university of applied sciences, financed by the Catholic Church, has been awarding bachelor's degrees in civil engineering since 2010. A master's degree is not offered but is planned. Despite what I consider to be quite high tuition fees of around 800 euros per year, around 300 young people, about 10 per cent of them from the neighbouring countries of Congo and Uganda, enrol here every year to study civil engineering. The teaching has to be handled by several lecturers with doctorates without a full professor. One of them is Francois Nsengiyumva, who welcomes me wearing a face mask and – today it is rainy and cool – a dark cardigan over a white shirt. The institute buildings, reinforced concrete frames with brick infill and blue tin roofs, are scattered across the well-kept, park-like campus. There is a well-equipped concrete laboratory here, and Francois would love to start joint research projects there right away. He is the contact person for SAICE-Africa, Capacity Building in Civil Engineering in Sub-Saharan Africa [5]. With INES as the centre, the aim is to create international

3

4

5

Figs. 3–5 Membrane roof at the Genocide Memorial in Kigali (3), and concrete-brick construction in the countryside and façade design by the local telephone provider (4–5)

6

7

Figs. 6–7 INES Campus Ruhengeri and the university's concrete laboratory

links between African universities and European partners in the field of civil engineering education. In November 2021, Francois organised an international conference on *Repositioning Civil Engineering Education for Sustainable Development* here in Ruhengeri, and the pride and joy he took in this is still evident during my visit six months later Figs. 6–7.

During my two visits to the universities in Kigali and Ruhengeri, I sensed a great interest in cooperation and enjoyed discussing it. Of course, I was concerned that I would return at the end of the trip with too many plans, but nevertheless I offered lectures via Teams, thought aloud about joint research projects, and discussed student exchanges in the form of internships with my colleagues. After the discussions, I suggested that these ideas be developed further and then sent to me. However, this did not happen. As long as I made no effort to stay in touch, my short visits were not enough to establish a lasting connection. Or did I overlook something? Did my colleagues perhaps assume that I owed them something?

Day 88, 10 June – Cyanika

An open-air fair within sight of the Ugandan border attracts attention with its small, pointed snow-white tent roofs. Around two dozen exhibition stands are neatly lined up around a rectangular square paved with concrete blocks. The local economy is on display, showcasing everything from wooden furniture from the village carpentry workshop to a small exhibition on healthy eating. I look at a table with a touching little collection of building materials. Are the bricks and tiles, sawn from hard, porous grey material, perhaps made of concrete? I can't clarify this with the representative at the table, so I buy one of her samples, which then makes the journey back to Berlin with us. Immediately after my purchase, I am cast by the local dairy for a milk advertisement Figs. 8–9.

In Berlin, Johannes Giebel, curator of the mineralogical collection at the Technische Universität Berlin, immediately recognises that my stone is of volcanic origin, a volcanite. The 39-year-old geologist, currently busy with lectures, three student interns, and his collection, travels widely around the world: on the TU website, he can be seen in expedition gear in Namibia. He is knowledgeable about rocks in Africa, but his interest has been piqued. So, he suggests that I subject my sample to a micro-XRF measurement – a detailed analysis of its chemical composition.

Johannes Giebel tells me that volcanic rocks such as basalt and plutonic rocks such as gabbro are magmatic rocks. I want to know more about this, and he takes the time to give me a short lecture: "Due to the lithostatic pressure inside the Earth, there is a higher 'gas' solubility in magma. When the lithostatic pressure is removed as the magma reaches the surface, the corresponding solubility drops abruptly and the 'gases' separate. This degassing process defines the transition from magma to lava. The rocks that form from magma through solidification while still inside the Earth are called plutonic rocks. The rocks that form from lava are called volcanic rocks. The two types of rock can be easily distinguished by their textures. Since plutonic rocks have significantly more time to crystallise in the Earth's interior (slow cooling), they often have a so-called holocrystalline structure (fully crystallised). Volcanic rocks, on the other hand, crystallise very quickly, and the lava is often even quenched by the significantly lower temperatures at the Earth's surface (compared to the Earth's interior). Therefore, volcanic rocks have a very fine matrix (groundmass). Bubble structures, which indicate degassing, are also characteristic of volcanic rocks. The slightly larger crystals in volcanic rocks, known as phenocrysts, crystallised as the magma rose towards the Earth's surface and were transported in the magma as it transitioned to lava until the volcanic rocks crystallised." The even more porous and very light pumice (→ Considerations – Concrete) is formed when lava is thrown into the air and the faster cooling process results in the formation of a porous aggregate of rock glass.

The evaluation of the micro-XRF measurement shows that I brought back a rock that is on the border between phonotephrite and tephriphonolite Figs. 10–11. Not unlike the phonolites I saw everywhere in Nairobi eight days after visiting the open-air fair in Cyanika! Such volcanic rocks characterise the East African Rift Valley from Ethiopia to Mozambique.

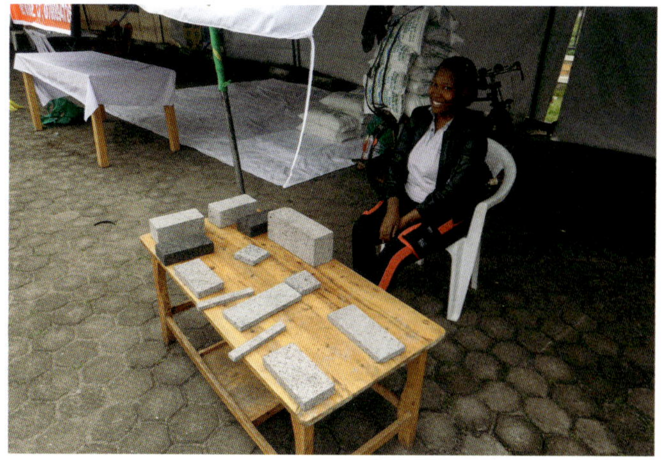

8

9

Figs. 8–9 Open-air fair in Cyanika: exhibition stand with building materials that could be concrete or of natural origin, and advertising for local dairy products (photo: Massimo Canossa; 9)

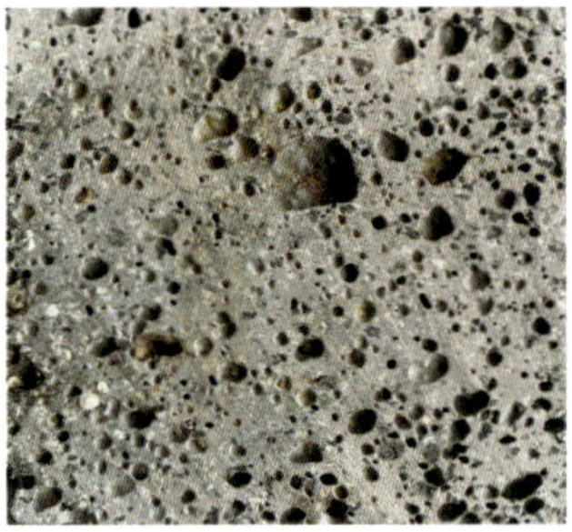

10

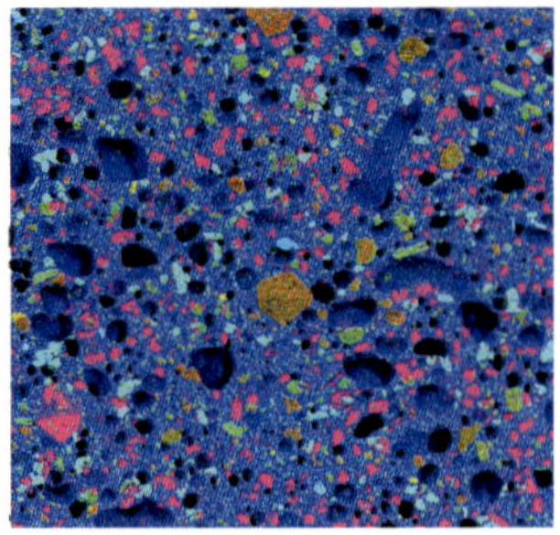

11

Figs. 10–11 Rock sample (front; 10) and micro-XRF element distribution map (back) with mineral composition (inclusions): orange = pyroxene, light green = amphibole, light blue = mica, pink = plagioclase (photo: Johannes Giebel; 11)

12

13

Figs. 12–13 A rock sample in the press (photo: Wilfried Walkowiak; 12), load-deformation curves of the two samples (13)

Can my phonotephrite be used for construction? We civil engineers can find out. At our Toni pressure testing machine at the Institute of Civil Engineering at the Technische Universität, we test two cubes that we cut out of the sample. The press slowly loads the samples with 0.6 N/mm²/sec until they break, measures the compression with inductive displacement transducers (WA), and records the load-deformation diagram Figs. 12–13. This reveals a modulus of elasticity of approximately 9,000 MPa, which corresponds to around one third of the stiffness of normal concrete. The test specimens only fail at a compressive strain of around 1 ‰ and compressive stresses in the order of 70 MPa. These are astonishingly high strengths for such a porous material. Building blocks made from this phonotephrite can be used to construct multi-storey buildings without any problems. Last but not least, we even find a short film showing how the exhibition exhibits were produced in Rwanda [6]: with their hands dangerously close to a huge stone saw blade, they are carved out of stone blocks piece by piece.

Training in Construction in Rwanda

Prof. Dr. Francois Nsengiyumva
Institut d'Enseignement Supérieur de Ruhengeri (INES), Musanze, Rwanda, July 2024

'Rwanda is a landlocked country located in eastern Central Africa. Its population is predominantly young, with approximately 70 per cent of citizens under the age of 30. This represents a significant potential workforce capable of driving socio-economic transformation, provided that young people are equipped with high-quality education and relevant skills. To address youth unemployment, the Rwandan government has prioritised education programmes that emphasise practical competencies. The government's plan is to enrol at least 60 per cent of upper secondary school pupils in Technical and Vocational Education and Training (TVET), the aim being to equip them with skills that are needed for entrepreneurship and job creation.

As in many developing countries, Rwanda is currently experiencing rapid infrastructure development, particularly in urban areas. Consequently, civil engineering has become one of the priority fields of study, attracting a large number of students. At the Institut d'Enseignement Supérieur de Ruhengeri (INES), for example, more than 10 per cent of all students are enrolled in the Department of Civil Engineering. At present, bachelor's degree programmes in civil engineering are offered by the University of Rwanda (public university), Rwanda Polytechnic (public university), and some private institutions including INES. However, postgraduate education in civil engineering remains underdeveloped. There is a notable shortage of experienced academic staff and well-equipped laboratories to support advanced training and research. In Rwanda, all new academic programmes must be accredited by the Higher Education Council (HEC) before implementation. The HEC, which operates under the Ministry of Education, is also responsible for evaluating and recognising foreign qualifications to ensure their equivalence within the national qualifications framework.

Rwanda still relies heavily on foreign consultants to design and implement major construction projects. This dependency highlights the urgent need to strengthen national capacity in engineering education, research, and professional practice.

The country currently lacks a railway network. Its mountainous terrain, characterised by steep slopes, presents considerable engineering challenges. Landslides are frequent, particularly in the western and northern regions, causing damage to infrastructure and, in some cases, loss of life. Despite these risks, effective mitigation measures are still limited. Furthermore, there is a scarcity of research literature on geotechnical engineering and landslide management in the Rwandan context.'

Addis Abba
ETHIOPIA
Hawassa
Yirga Alem
SOUTH SUDAN
Konso
Turmi
Kakuma
Lake
Turkana
Bidi Bidi
Atiak
Patiko
Gulu
DEMOCRATIC
REPUBLIC
OF THE CONGO
UGANDA
KENYA
Namugongo
Kamuli
Mbale
Day 89, Kampala
Kampala
Day 92, Jinja
Samburu
Mityana
Entebbe
Miwani
Kyambura
Njoro
Mannya
Kabale
Nairobi
Mlolongo
Gihembe
Kayonza
RWANDA
Kigali
Rwinkwayu
Emali
Shela
Lamu
Bukavu
Mwanza
Huye
Moshi
BURUNDI
Arusha
Mombasa
TANZANIA
Pangani
Dodoma
Bagamoyo
Dar es Salaam
0 km
250 km
Mikindani

Uganda
10–13 June 2022

The East Africa Tourist Visa (EAC), introduced in 2014, which we obtained for Rwanda, is also valid for Kenya and Uganda and makes our entry today easy. Unfortunately, we don't have much time for the country. Massimo is ill, the cracked windscreen can only be replaced in Kenya, and our wives will soon be arriving there, too. Paloma and I are going to take a two-week 'holiday' and make a detour back to Tanzania. That's why there are two travel periods in the diary for Kenya and Tanzania. This makes what I remember from the four days in Uganda all the more astonishing.

We spend the first night at a campsite that we have all to ourselves thanks to the Covid-19 pandemic. It's right on the shore of Lake Bunyonyi. There, a sign warns non-swimmers about "Uganda's deepest lake", which is up to 2,000 metres deep: "The water is very deep. You can drown with ease!" According to the Lonely Planet travel guide, Lake Bunyonyi is "undoubtedly the loveliest lake in Uganda". We reach it via a dirt road, alongside which I notice quarries where stone is being mined by hand. The men chop blocks out of the mountain, and below, next to the road, women and children squat and break them into small stones with hammers. These can be used as aggregates for concrete. In fact, this scene, which is literally from the Stone Age, remains more vivid in my memory than the romantic atmosphere at the lake between the steep green mountain slopes.

1

2

Figs. 1–2
Lake Bunyonyi and quarry on the way there

Day 89, 11 June – Kampala

Wait, what's going on here? Why are there so many stalls selling colourful T-shirts and soft drinks in the middle of nowhere, about 70 kilometres from Uganda's capital Kampala? We almost drove past the equator! It is marked by a line between two inconspicuous, white-painted, man-sized concrete rings on either side of the road, and we have now covered almost exactly half the distance between Cape Town and Cairo Fig. 3. We dutifully queue up in the blazing sun, humid air, and temperatures of over 30° Celsius until we can climb into one of the rings and take our photo as proof.

Kampala awaits us with a maze of streets that even Google Maps has trouble navigating, traffic jams full of lorries, and police officers who, before showing us the way, give us a ticket for crossing a solid line to pull up alongside them. It is hard to imagine that, according to forecasts, this city will grow from its current population of two million to 40 million by 2100, making it one of the 20 largest cities in the world [1] Fig 4.

Why isn't Makerere University responding to my emails? I go there anyway because this university, founded in 1922, is one of the most important universities in Africa and has more than 30,000 students. Well-known personalities such as Julius Nyerere and Milton Obote, Uganda's first prime minister, studied there. [2] In the centre of the park-like campus, I first look for the face of the university, the central administration building with its striking 'Ivory Tower' built in the 1930s – and instead I find a construction site. The building was heavily damaged by a fire in September 2020 – "a disaster for Uganda and for East Africa" [3] because the building's archives contained files on the entire history of the university. The four-storey brick building surrounded by trees, which houses the civil engineers, welcomes me with a sign that I would also like to see at German universities: "College of Engineering, Design, Art, and Technology". How good it is to see engineering embedded in art and design. On site, I learn that the email addresses on the university's website are out of date. Later, however, the principal, Professor Henry M. Alinaitwe, who is planning the reconstruction of the 'Ivory Tower', will write the text at the end of this chapter Figs. 5–6.

3

4

Figs. 3–4 Across the equator near Kampala and acrobats in the city's traffic

5

6

Figs. 5–6 The fire-damaged central administration building and the College of Engineering, Design, Art and Technology at Makerere University

7

8

Figs. 7–8 The Source of the Nile cable-stayed bridge near Jinja

Day 92, 13 June – Jinja

From Jinja on Lake Victoria, the White Nile flows almost 3,000 kilometres to Khartoum, where we will see it in eight weeks. Here the river is still called the Victoria Nile, and to reach the city from Kampala, we have to cross it on the Source of the Nile Bridge. The Kampala–Jinja Highway, the A109, which will take us there, is full of trucks transporting overseas containers, honking minibuses, and mopeds slaloming between them. Concrete slabs have been laid over a deep drainage ditch that opens up right next to our wheels, providing refuge for pedestrians scurrying across the road. This takes them to snack bars with colourful plastic chairs and tables, open-air moped repair shops on oil-soaked earth, yellow-painted telecom shops, and a construction equipment rental company with rusty concrete mixers and road rollers. Alongside this bustle, we drift down the river of vehicles to the Nile and can already see from a distance the 100-metre-high bridge pylons with their white cables standing like sails in the landscape. The buildings become sparser and the traffic jam clears in front of the 525-metre-long cable-stayed bridge with a main span of 290 metres. The two pylons and the stay cables support a centrally suspended, 22.9-metre-wide, and only 2.5-metre-high three-cell concrete box girder for a four-lane road with hard shoulders. Bridges are critical infrastructure, and it takes some persuasion on my part before the uniformed and armed security guards allow me – "a civil engineer, who has travelled here specially from Germany" – to take close-up photos of this elegant structure.

According to the plaque next to the bridge, the Source of the Nile Bridge was inaugurated in October 2018 by President Yoweri Museveni "as a token of friendship and collaboration between Japan and the Republic of Uganda". The bridge, financed by Japan, was built by the Japanese construction company Zenitaka in collaboration with Hyundai Engineering and Construction from South Korea. Planning and construction supervision were the responsibility of Oriental Consultants from Tokyo Figs. 7–8.

Jinja has another surprise in store for me: I am delighted to discover some architectural gems. Decaying modernist villas on the shores of Lake Victoria and Indian colonial buildings such as the Madhvani Building from 1919, the city's first multi-storey building, on Main Street invite me to explore. There is very little information available about these enchanted buildings in various stages of decay and, in some cases, renovation, but an internet search quickly reveals that this barn find is not unknown [4] Figs. 9–10.

Indians, Pakistanis, and Bangladeshis, who had come to the country in the wake of the British and the construction of the railway in the 1920s, had obviously achieved a degree of prosperity here as traders, mechanics, doctors, and teachers. That is until 50 years ago when Idi Amin Dada, "the Butcher of Uganda", forced them to leave the country within three months. [5] With a divine inspiration in a dream – "Asians came to Uganda to build railways. The railway is finished. They must leave now." [6] – the dictator justified the cruel and degrading expulsion of up to 70,000 people in the summer of 1972. In doing so, Idi Amin finally drove the country's already weak economy towards collapse.

Today, Uganda seems to have made peace with this terrible chapter in the country's history. The workshop where our Land Cruiser with a cracked windscreen gets an initial inspection is owned by Karan, an Indian, and we hear that Indian factories have now re-established themselves here. Jay has also recently arrived from India and already knows that he wants to stay here. He runs the small Forever Resort where we are staying, which belongs to his uncle. Mothers dressed in colourful saris let their children play here on the lawn right by the Nile, and the chef from Nepal barbecues fresh Nile perch for us.

9

10

Figs. 9–10 Modern-style villa and the Madhvani Building in Jinja

Civil Engineering in Uganda

Prof. Henry Mwanaki Alinaitwe,
Makerere University, Kampala, Uganda, August 2024

'In order to practise as an engineer in Uganda, you must be registered with the Engineers Registration Board (ERB; Act Cap 271). Foreigners who want to work legally as civil engineers here must register temporarily. The Engineers Registration Board lists a total of 1,226 civil engineers. In fact, civil engineers make up the majority of registered engineers. Uganda has agreed on mutual recognition of engineers with other East African countries. A recognised degree in engineering and at least four years of professional experience under the supervision of registered engineers are currently the requirements for registration with the ERB. In addition, candidates must submit a report on their professional career and a technical report, which will be reviewed.

Civil engineering at bachelor's level was introduced at Makerere University in 1970 as an elective subject in engineering alongside mechanical and electrical engineering. From the mid-1990s onwards, students were able to enrol separately for civil engineering, mechanical engineering, or electrical engineering. There are currently more than six other universities in Uganda offering bachelor's degrees in civil engineering. Graduate programmes in civil engineering are currently offered at Makerere University and a few other universities.

Before 1970, all engineering graduates had been trained outside Uganda, mainly in Nairobi, Kenya. The oldest generation of engineers had a diploma-level education. However, they were certified by City & Guilds of the United Kingdom, which enabled them to practise their profession. Most of the older civil engineers with a master's or doctoral degree obtained it outside Uganda, mainly in the United Kingdom and the United States.

Civil engineers in Uganda cover all areas of construction. They work mainly in the fields of building construction (public and private, residential and commercial buildings, office and industrial buildings, leisure and religious buildings, schools and hospitals) and road, bridge, and rail construction. They design dams, water-works, oil fields, pipelines, airports, and port facilities. Thanks to the country's political stability, the construction industry has been able to expand over the past 40 years. Following the civil wars of the 1980s and 1990s, the volume of construction has increased rapidly.

In terms of road construction, there is a distinction between paved national roads and unpaved country roads. The government is in the process of increasing the proportion of paved roads. Local materials such as reeds, grass, clay, and wattle, as well as wood or bush poles, are used to build houses in rural areas. Although these buildings are not very durable, they are well suited to local climatic conditions due to their affordability, good ventilation, and thermal insulation. Thatched houses are widespread in many regions of Uganda, although their proportion is steadily declining.

Uganda has many lakes, rivers, and streams, and the source of the Nile is located in Jinja. Water supply systems are numerous and largely decentralised. Cities have many wastewater treatment systems, sewage treatment plants, and supply systems. Their planning and construction account for a large part of today's engineering activities.

Large hydroelectric power stations, port facilities on Lake Victoria, bridges, airports, sports stadiums, and other structures are currently being built in Uganda, creating opportunities for innovative designs using modern materials. Reinforced concrete remains the predominant building material.

There are many small and medium-sized companies in the industry that carry out the majority of construction projects. However, most of the large-scale construction projects are carried out by international engineering firms and construction companies. These receive support from their countries of origin in the form of better financing and easier procurement of equipment. Added to this is their many years of experience. This creates the risk that they will squeeze out the small local companies. Nevertheless, the construction industry in Uganda is growing.

The future of civil engineering in Uganda looks bright. Due to the enormous demand for more and better roads, standard-gauge railways, water supply and sewage disposal for all communities, secure energy supply, more housing for the growing population, and also because of the discovery of oil reserves, the tasks are constantly increasing. Uganda needs to train and qualify more civil engineers and provide funding at reasonable interest rates in order to create a better construction infrastructure.'

Bahir Dar
Dire Dawa
Harar
Addis Ababa
ETHIOPIA
Hawassa
Yirga Alem
SOUTH SUDAN
Konso
Turmi
Kakuma
Lake Turkana
SOMALIA
Bidi Bidi
Atiak
Patiko
Gulu
Moyale
UGANDA
KENYA
Namugongo
Luweero
Kamuli
Mbale
Kampala
Jinja
Mityana
Samburu
Miwani
Kyambura
Entebbe
Njoro
Day 96+119, Nairobi
Nairobi
Kabale
Mannya
Gihembe
RWANDA
Kayonza
Rwinkwayu
Kigali
Mlolongo
Emali
Shela
Lamu
Bukavu
Huye
Mwanza
Moshi
BURUNDI
Arusha
Day 110, Mombasa
TANZANIA
Pangani
Dodoma
Bagamoyo
Dar es Salaam
0 km
250 km
Kilwa Kisiwani
Songo Mnara

Kenya via Nairobi to Tanzania 13–24 June and from Tanzania via Nairobi to the north 1–15 July 2022

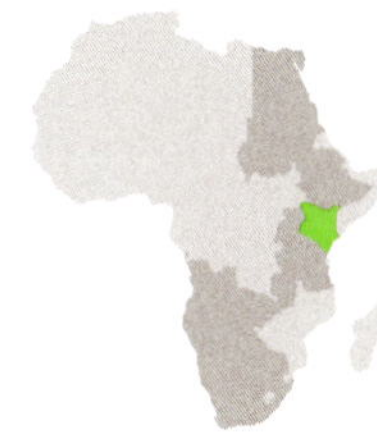

From rainy Busia on the western border with Uganda, past the huge tea plantations near Kericho, which lie like green carpets on the slopes of the Great African Rift Valley, into the exciting metropolis of Nairobi and then on to the estate of German-English friends at the foot of Mount Kenya: It feels as if I am stumbling from one scene to the next in the film *Out of Africa*. We continue to the tropical green picture-perfect coast of Malindi and back through the dusty national parks of Tsavo and Amboseli to Masai Mara, where over a million wildebeests are currently migrating. A month later, we cross the Dida-Galgalu desert heading north towards Ethiopia. A few years ago, the road to the north was notorious for regular attacks on travellers, but we drive safely and quickly on a new asphalt road to the border at Moyale.

Kenya, a developed and fascinating travel destination, is on the eve of elections. Large posters are everywhere, and the green and yellow ones of the smiling candidate William Ruto are particularly large and conspicuous. The last election was marred by bloody unrest, so we are glad to be able to leave the country before the general election on 9 August. Despite some commotion on election night, the elections will proceed calmly overall. After the Supreme Court subsequently confirms the elections, William Ruto, previously Vice-Chancellor under Uhuru Kenyatta, will become the country's fifth president on 13 September 2022 and continue the pro-Western policies of his predecessors.

Kenya is a fairly safe country in which to travel, although we do not walk around at night in big cities like Nairobi and Mombasa. And, as in the rest of Africa, we try not to drive on country roads after dark. We encounter police checks surprisingly often, and it is not uncommon for us to be politely asked what a euro coin looks like. Most of the time, we get away with a friendly smile and a cheerful "next time", but I can imagine that annoyances like this make everyday life in Kenya more difficult.

Day 96, 18 June – Nairobi

Nairobi, the capital of Kenya since 1905 (at that time 10,000 people lived in the city; today there are 4.5 million), is also the phonolite capital. This volcanic rock, with its grey to yellowish colour and light-coloured feldspar inclusions, shapes the face of Nairobi Figs. 1–3. The city and its surroundings lie on a 30- to 40-metre-thick layer of phonolite [1] close to the surface, and this inexpensive local building material is used extensively. On the way northeast of Nairobi towards Jomo Kenyatta University of Agriculture and Technology (JKUAT), I see the quarries in Juja where the grey variety of the stone is mined. As the name suggests, the stone that sounds when struck can be used for lithophones. Here, however, it is cut using imperial measurements for use in building construction. The 15-inch-long and 6-inch-high phonolite blocks are available in widths of 4, 6 and 9 inches, and we see them everywhere in the city as masonry that does not need to be plastered. In single-storey buildings, the walls are made entirely of phonolite, and in multi-storey buildings, they are strengthened with reinforced concrete frames.

1

2

3

Figs. 1–3 Houses made of unplastered phonolite masonry characterise Nairobi's cityscape

I gather from the literature that the compressive strength of phonolite ranges between 15 and 100 MPa, which is similar to that of concrete. [2] Because the stone is quite porous, I imagine that it has better thermal insulation properties than concrete. This is certainly true of another, even lighter volcanic rock in the area: Kerichwa Valley tuff, which is also used in residential construction. [3]

Phonolite is so strong that it can also be used as an aggregate for concrete when crushed. For the 27-kilometre-long Nairobi Expressway, the responsible Chinese construction companies were able to use phonolite aggregates to produce concrete with strengths suitable for bridge construction (C30 and C60) as a substitute for rare river gravel and sand. [4] When we arrive in Nairobi, this toll road has just been inaugurated. We now reach Jomo Kenyatta Airport from the city centre in 20 minutes, driving comfortably on an elevated road. Until April 2022, this journey took two hours. The China Road and Bridge Corporation built this structure in an incredible 21 months. This is what Simon Miring'u from the Kenyan branch of SMEC tells me. The young, smart engineer, dressed casually in a blue shirt, beige cotton trousers, and freshly polished leather shoes will take me to 'his' JKUAT a few days later and patiently answer my questions again and again, even after the trip. He and his colleagues were responsible for supervising the construction of the Nairobi Expressway, and he shows me the 'construction site'. Simon is convinced that only 'the Chinese' can build so quickly, and when I think of Germany and our bridges, I can only agree. With all due respect for this achievement, however, when I look at the bridge, I cannot help but think that there has been at least one casualty here: *Baukultur* Figs. 4–5. Simon tells me that quite a few civil engineers from Ethiopia work in Kenya. A work permit for foreigners costs around 5,000 US dollars, and the additional temporary engineering practising licence costs another 5,000. With this expensive hurdle, Kenya ensures that only highly qualified workers who are worth the investment or those who can afford it come to the country.

4

5

Figs. 4–5 On and under the Nairobi Expressway

6

7

8

9

Figs. 6–9 Kenyatta International Convention Centre in Nairobi's Central Business District

The buildings in Nairobi's Central Business District (CBD) bear witness to the spirit of optimism that prevailed after the country gained independence in 1963. The moderate policies of the first Prime Minister, Jomo Kenyatta, attracted international investment and filled the CBD with United Nations buildings, a convention centre, hotels, the City College, and offices. After that, the focus shifted, and the large buildings of recent decades are located elsewhere. Time seems to stand still in the CBD, and we stroll (during the day) through an open-air museum of 1960s and 1970s architecture with beautiful exposed concrete buildings, such as the Cathedral Basilica of the Holy Family (1960), the concrete shell roofs of Wakulima Market (1966), and, of course, the Kenyatta International Convention Centre (1973), an exposed concrete masterpiece whose tower was the tallest building in East Africa for a long time. The ticket for the observation platform is a good investment, rewarded with a spectacular panoramic view of the city centre. After the lift ride, we reach the roof via the abandoned restaurant below. All that remains of the restaurant are the parquet floors and bare walls, and no one prevents us from exploring this enchanted two-storey space – via free-standing concrete stairs – with distant views through dusty windows and bringing it back to life in our minds Figs. 6–9.

Parabolic concrete arches with horizontal window bands and concrete beams to stiffen them form the roof of the light-flooded City Market, which was built in the early 1930s Figs. 10–12. The many stalls on two floors are geared towards tourists. Anyone who comes here and says they only want to take photos should not expect the many enthusiastic souvenir sellers to leave them alone. The City Market appears to be a replica of Lawrence Hall in London, completed in 1929, whose supporting structure was in turn influenced by the French concrete pioneers François Hennebique and Eugène Freyssinet [5]. The architecture of the CBD is described in detail in the two books recommended in the introduction (see also [12–13]).

10

11

Figs. 10–11 City Market in Nairobi, the interior and the reinforced concrete supporting structure

12

Fig. 12 Colourful façade of City Market, Nairobi

Day 110, 2 July – Mombasa

National Highway B3 leads from Masai Mara, Kenya's most beautiful national park in the low-lying Great African Rift Valley, up to Nairobi. Due to the many trucks driving uphill at walking pace and racing downhill, the journey is already a test of nerves. Anyone wishing to reach Kenya's second largest city, Mombasa, located on an island, from the south coast via the A14 international highway must ultimately abandon the road altogether and battle their way through a long, multi-lane traffic jam to the Likoni ferry, which has been the only connection to the south for 85 years Fig. 13. Making this crossing once can be considered an adventure, but for those who have to cross here every day, it is hoped that the construction of the Mombasa Gate Bridge, financed by Japan and postponed several times, will begin soon [6]. The cable-stayed bridge, with a main span of 660 metres and two side spans of 330 metres each is reminiscent of the Sunshine Skyway Bridge in Florida and would then surpass the Msikaba Bridge in South Africa as the largest cable-stayed bridge in Africa and end the suffering of ferry users. However, around 12 kilometres of roads and access ramps still need to be built to reach the elevated bridge. It must cross the water at an impressive height of around 70 metres so that container ships can pass underneath it to Kenya's largest port.

The A109 expressway from Mombasa back to Nairobi is well developed, but here, too, heavy goods traffic is a constant threat to its peers and to small cars. In 1983, I was travelling parallel to this road, which crosses Tsavo East National Park, on a metre-gauge railway line in a train from the colonial era. It was a slow and dreamlike journey with sleeping cars and morning tea served in silver-plated pots, passing giraffes and elephants. Since 2017, the new standard gauge railway line, the Mombasa–Nairobi Standard Gauge Railway (SGR), has been in operation here, providing a modern and much more efficient connection between Mombasa and Nairobi for the Kenyan economy. As part of this new construction, around 100 bridges were built [7], the longest of which is the Athi River Super Bridge, opened in 2017, at 2.6 kilometres [8]. Like all bridges on this line, it consists of a long series of 30-metre-long precast girders with an estimated height of 3 metres. These are reminiscent of the 40-metre-long single-span girders in the Deutsche Bahn's framework planning. It is a pity that more time was not taken for planning and construction to build integral bridges. Why is it that, just like in Germany, the realisation is only slowly gaining ground that bridges without joints and bearings are not only more material-efficient, durable, and inexpensive than single-span girders, but also slimmer and – in my opinion – more attractive Fig. 14?

13

Fig. 13 Overloaded Likoni ferry in Mombasa

14

Fig. 14 One of the prestressed concrete bridges on the railway line from Mombasa to Nairobi along the A109

Day 119, 11 July – Nairobi 2

I am told that the best place to study civil engineering in Kenya is at the University of Nairobi. It is also possible to study at Jomo Kenyatta University of Agriculture and Technology (JKUAT) or Kenyatta University in Nairobi, and Moi University in Eldoret, around 300 kilometres north-west of Nairobi, is also mentioned. Professor Stanley Muse Shitote explains to me on the phone that Moi University, the second oldest state university in the country, offers a five-year bachelor's degree in civil engineering. The focus is on an undergraduate programme with around 50 students per year. A full professor, two associate professors, and 12 staff members are available for this purpose. Master's programmes are not offered, and the focus of research is on concrete and concrete renovation as well as affordable housing.

Of course, I would like to visit the University of Nairobi, but on Friday, the following Monday, when my visit is planned, is declared a "public holiday, marking Iddul-Adha" (the Islamic Festival of Sacrifice). I can't reschedule at such short notice, so my visit is cancelled. At least I managed to visit Jomo Kenyatta University of Agriculture and Technology beforehand. It is located on a sprawling green campus just outside the city. Simon Miring'u, who studied there, takes me. Dr. Mung'athia, the Head of Department, a gaunt, somewhat taciturn but very friendly figure of authority with a large JKUAT pin on his lapel, takes me to a lecture – I have the feeling it's the nearest one available – interrupts it and lets me talk a little about bridge building. No one seems to mind. Afterwards, he leads his former student and I across the campus to the well-equipped large test hall where all machinery needed for classic concrete research, such as testing concrete cubes and beams, can be found. In the laboratory, civil engineers are experimenting with sustainable materials. I see masonry made of cement-strengthened earth and coconut fibres, used to reinforce such bricks. Of course, concrete and steel are also the most important building materials in Kenya, and the country can produce its own cement. However, steel still has to be imported Figs. 15–17.

15

16

17

Figs. 15–17 The campus of Jomo Kenyatta University, with Simon Miring'u and Dr. Mung'athia in the concrete laboratory and materials for sustainable construction

18

Fig. 18 A footbridge near Nairobi – only the truss carries the loads …

Here, at the College of Engineering and Technology, which includes the School of Civil Engineering, Environment, and Geomatics with the Department of Civil Engineering, a five-year bachelor's degree programme focuses primarily on building and road construction, hydraulic engineering, and environmental engineering. In the fifth year, students specialise and can then take extra units in structural engineering, for example. One unit corresponds to six hours per week, with four hours of lectures and two hours of practical work. Eight such units must be attended in the first semester alone, which means 48(!) hours per week. Jomo Kenyatta University of Agriculture and Technology started as a college in 1981 and has been an independent university since 1994. Every year, 200 students begin studying civil and structural engineering, with women accounting for 25 to 30 per cent of the student body. The master's programme further specialises in building and road construction. There are currently around 20 students in the master's programme and around 20 PhD students. The department has two full professors, four associate professors, and 12 senior lecturers with PhDs.

Kabushiya
Naqa
Omdurman
Khartoum
Kassala
ERITREA
SUDAN
Aksum
Gheralta
Wukro
Mekelle
Day 137, Gondar
Day 134, Lalibela
Logiya
DJIBOUTI
Bahir Dar
Dire Dawa
Harar
Addis Ababa
Day 130+131, Addis Ababa
ETHIOPIA
Day 127, Shashemene
Hawassa
Yirga Alem
SOUTH SUDAN
Konso
Day 125, Turmi
SOMALIA
Kakuma
Lake Turkana
Atiak
Patiko
Gulu
UGANDA
KENYA
Namugongo
Kamuli
Mbale
Kampala
Samburu
Miwani
Njoro
0 km
250 km

Ethiopia
15–31 July 2022

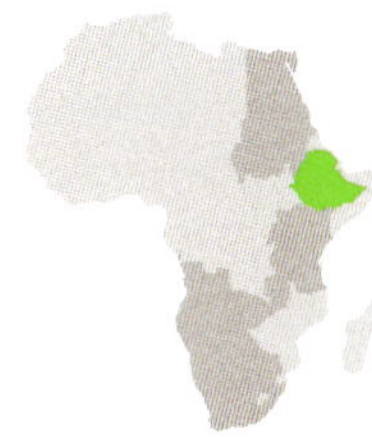

Ethiopia is the country richest in sights and the most contradictory of those we have visited on our journey so far. The Hamar people in the remote southwest of the country, with their archaic rites, seem to come from a bygone era. A few hours' drive further on, we stop at a solar kiosk from Berlin. In the north, not far from the Danakil Depression, which is one of the hottest deserts in the world and is at 100 metres below sea level, mountains over 4,000 metres high rise up. This highland is the source of the Blue Nile, which supplies around 80 per cent of the Nile's water that eventually reaches the Mediterranean Sea. A proud nation with a culture dating back thousands of years, virtually untouched by colonial influences, coupled with strong economic growth – it was even the highest in the world at 10.2 per cent in 2015 [1] – has raised hopes that are currently being thoroughly shaken.

In July 2022, we see a country plagued by armed conflict. Travelling through Ethiopia is, therefore, not easy. There is unrest practically everywhere outside the capital: in the Southern Nations, Nationalities and Peoples' Region, in Oromia, and on the border with Sudan. We are fortunate and are always in these places when it is quiet. We do not have any problems obtaining a tourist visa and, apart from constant checks by the police, soldiers, and militias, we have no difficulties.

There is war in the far north in the Tigray region on the border with Eritrea. Until 2018, the Tigrayans had dominated Ethiopian politics. Then they were replaced by Abiy Ahmed, an Oromo who received the Nobel Peace Prize in 2019. In 2020, with support from Eritrea and the Fano militias from the Amhara region, his government troops pushed rebellious Tigrayans, who had advanced far to the south, back to Tigray and are now starving them and the population. A ceasefire is finally negotiated in November 2022. The writer Navid Kermani, who is travelling in Tigray and Amhara just a few months later, recalls on Deutschlandfunk radio and in the weekly *Die Zeit* newspaper the "forgotten war" that has cost the lives of several hundred thousand people within two years. [2] In August 2023, conflict breaks out between the former allies. In Amhara, people feel that the government is not providing sufficient protection against Tigray, and the Fano fighters, therefore, refuse to surrender their weapons. Government troops are now also fighting in Amhara. [3] In the capital Addis Ababa, people feel they are in the right and misunderstood by the Western world. During our stay there, I see stickers from the #NoMore movement – blocked by Twitter – everywhere I go: "Its central objective is to oppose the ongoing Western media disinformation campaign, Western economic warfare, diplomatic propaganda, and military interventions in Africa in general, and the 'Horn of Africa' in particular." [4]

The Ethiopians are so proud, proud of their own history, proud of not having been colonised – and yet their country is dependent on donors. According to the World Bank, foreign aid, especially from the United States, accounts for around half of Ethiopia's gross domestic product. [5] The population is growing rapidly and has quadrupled since 1950 to around 120 million people in 2022. I learn that the forest area, which 60 years ago still accounted for 40 per cent of the country, has now shrunk to around 15 per cent. Only half of the population has access to electricity, and there is a widespread lack of clean water and drainage. Even in the capital Addis Ababa, only slightly more than 10 per cent of residents have toilets connected to the sewage system. [6]

The Grand Ethiopian Renaissance Dam (GERD), which dams the Blue Nile about 10 kilometres from the border with Sudan, is therefore very important for the country's development. The connected hydroelectric power plant, which is slowly being connected to the grid, is set to become the largest in Africa with a capacity of 6,000 MW. At the same time, the reservoir is a major bone of contention with Sudan and Egypt, countries that have themselves dammed the Nile in many places, with all the associated side effects. From what I hear from colleagues there, this dam, which is so important for Ethiopia, should not have a major impact on the water supply of Sudan and Egypt, provided that there is good coordination. Even evaporation, which leads to noticeable water losses in the hot countries of Egypt and Sudan, should only play a minor role in the cooler and rainier north-west of Ethiopia.

There is said to be little Chinese influence on the GERD, which is otherwise visible everywhere. The smelly old Lada taxis in Addis Ababa are being converted to LNG/electric, and the cars come from China. Of the country's 16 cement factories, China built the last six. I hear comments like: "Europe is forgetting us"; "A German engineer costs 20,000 euros per month, a Chinese engineer only needs food, cigarettes, and clothing"; "To get money from the World Bank (USA) or the International Monetary Fund (Europe), you have to subordinate yourself"; "If you want money and technology, you can only get it from China at low interest rates"; "The Chinese don't ask questions about democracy, human rights, or corruption"; "The Chinese are different. They observe the market and flood it with their products". I understand that in times of great need, you take money from those who ask the fewest questions. But you still have to pay it back.

Day 125, 17 July – Turmi

The unrest throughout the country and the travel warnings we read on the German Foreign Office website seem to have put an end to tourism as an important source of income in 2022. Apart from a few embassy staff and NGO workers travelling around the country, we see hardly any tourists during our stay. We want to visit the Hamar people (also known as Hamer or Amar) and we have the villages on the border with South Sudan all to ourselves According to Lonely Planet in 2017, they were "often bursting at the seams with tour groups". I could indulge in the illusion of undiscovered land, were it not for the bitter aftertaste of knowing the circumstances. However, it seems that tourism here has moved beyond the phase of 'human safaris' and is offering local people the opportunity to profitably preserve their unique cultural assets for the future. Road construction to open up this remote part of the country is continuing, with routes now accessible not only to sturdy off-road vehicles. Everywhere along the dusty track from Weyto via Arbore to Turmi, which is difficult to pass during the rainy season, we see road construction work. Soon, asphalt roads will expose the indigenous traditions of the region's ethnic groups even more to 'civilisation', and 'respectful tourism' could become a stabilising factor there. [7]

In Turmi, we find a telephone number at the locked office of the Local Tour Guide Association and we use it to reach Asu. The man in his mid-twenties from the Hamar tribe is armed with a mobile phone, dressed in jogging pants, a polo shirt, and sneakers, and speaks reasonable English. He offers to take us to a Hamar village where we can also spend the night. The phone rings, and he seems genuinely surprised when he learns that a Hamar wedding, a so-called bull jumping ceremony, is to take place in a few hours. We postpone our visit to the village until the evening and set off on an adventurous excursion. Shortly after Turmi, we turn onto a road that hasn't seen a car in a long time. It becomes increasingly narrow and, after an hour's drive over rough and smooth terrain, it ends in a riverbed. We continue for a few kilometres on its gravel bed lined with scrub, and just as I am about to turn back because the journey is starting to feel eerie, Asu tells us to stop. We walk a few hundred metres through the bushes on dry ground and suddenly find ourselves among around 200 Hamar who are in the process of lining up several cows and oxen in rows. Asu briefly introduces us to the village elder, and after paying a small fee, we are allowed to stay. I don't think we're disturbing anyone; they simply ignore us. The bull jump is an initiation rite in which a future groom must jump naked onto the lined-up cattle and balance himself on them. If he manages to do this several times back and forth, he is allowed to marry. I have never seen people dressed and made up in such creative and colourful ways. At first, I mistake the white ornaments painted on the men's bare legs for knee socks. Colourful cloths are knotted into skirts, capes, and headgear. The Hamar have turned copper rings and coloured beads into extravagant body jewellery with nails, telephone cards, and stainless-steel watch straps. In addition, almost every adult man carries a Kalashnikov (available here for 800 to 1,000 US dollars) with a cartridge belt (2 dollars per cartridge). The men's headdresses are elaborate clay amulets glued to their hair, into which an ostrich feather can be inserted. In the past, this was used to show that the wearer had recently killed a dangerous animal or enemy. These warriors carry a borkoto in their hands, a small headrest used at night to preserve the headdress – now only a decorative element – for several months. The women, who have shaped their hair into copper-coloured braids using earth, paint, and resin, dance and sing in groups while today's candidate concentrates on the jump. The big moment arrives, and the future groom completes the strenuous exercise. Less than ten minutes later, the event is over and we stand alone in the clearing between the bushes. The families still have to walk up to 20 kilometres back to their farmsteads. Completely dazed, we drive back to Turmi Figs. 1–3.

1

Figs. 1–3 Hamar wedding – the bull jumping ceremony

The approximately 50,000 Hamar are one of the 16 ethnic groups in the region, and they "have always practised a mixed subsistence economy based on slash-and-burn agriculture, supplemented by raids" [7]. Their lives revolve around farmsteads, which are usually located on high ground for a better overview and consist of several houses with cattle pens and goat enclosures. The walls of the circular houses are made of wooden poles as thick as a person's wrist, embedded in the ground and held together at the top by a ring of additional poles. This ring also supports the cone-shaped roof, which is made of wooden poles as well and covered with grass. The roof edge, which protrudes about one metre beyond the wall, is supported by thin posts and forms a protective veranda around the perimeter. The walls are covered with clay on the inside up to just below the ring. The fireplace inside, which serves as the kitchen and sole source of lighting, allows smoke to escape outside through the gaps above it. We are served a drink made from coffee bean husks – we brought a large bag of them with us. It bears only a faint resemblance to 'caffè crema' and is drunk from a pumpkin shell. Clothes and tools hang on the walls, all made of wood, iron, leather, or grass, except for the sandals, which, as is often the case in Africa, are made from old car or bicycle tyres. Although it gets quite cool at night, when we are there, all the inhabitants sleep in front of the house, wrapped in cloths on goat and

2

3

cattle skins. Presumably, it is too smoky inside the house. Much of it reminds me of the similar huts and equally sparse possessions of the Maasai, whose bomas we visited about a month ago less than 1,000 kilometres south as the crow flies. We are tired and retire early to the car parked nearby. The Hamar walk and sit around us until late into the night, and the sounds of the village lull us to sleep. The next morning, Asu wakes us up and we step into an empty village. The Hamar have already set off for work Figs. 4–11.

Before we leave Turmi, we drink another coffee at Graft's solar kiosk. The Berlin architects developed a PV-equipped modular system consisting of aluminium frames and filling elements, which they installed in over 200 locations in Africa. In villages without electricity, these solar-powered kiosks provide light, chilled drinks, and mobile phone charging. [8] The solar kiosk here also initially produced electricity and did what it was supposed to do successfully. Turmi is now connected to the electricity grid and the solar kiosk has been converted into a café with a bar.

4

5

6

8

7

Figs. 4–11 Hamar houses and their inhabitants

9

10

11

12

13

14

15

Figs. 12–13 Rastafarians in Shashemene and reggae souvenirs in Addis Ababa

Figs. 14–15 Busy thoroughfare and expressway towards Addis Ababa with a frame bridge as an overpass

Day 127, 19 July – Shashemene

On the way to Addis Ababa, we reach Shashemene, the 'capital' of the Rastafarians. The followers of this Christian-oriented religious movement, which originated in Jamaica in the 1930s, are concerned with the liberation of black people. The Rastafarians are named after Emperor Haile Selassie (1892–1975) – whose name was Ras Tafari Makonnen before his coronation in 1930 – because a black king in Africa had been prophesied to them. When the emperor allocated 200 hectares of land near Shashemene to black immigrants in 1950, Rastafarians from all over the world settled there to be close to the reborn Messiah. We stay at the Rastafari Hotel Zion Train and hear from our hosts that there is no evidence of Haile Selassie's death and that the apocalypse is near. In this respect, the Rastafarians remind me of the Temple Society, the deeply religious Swabian Pietists who settled in Palestine at the end of the nineteenth century to experience the expected end of the world in the Holy Land. The unusual settlers in Shashemene are still not well integrated today. They are repeatedly threatened with expulsion. In Addis Ababa, on the other hand, the movement, reduced to Bob Marley and ganja, can be seen in the form of green, yellow, and red souvenirs at every turn Figs. 12–13.

The further north we drive towards the capital, the better and busier the roads become. Cars, buses, tuk-tuks, and motorcycles share the road with ox and donkey carts as well as many people, sheep, and goats. With potholes and frayed roadsides adding to the mix, the stretch of the A7 from Arba Minch to Ziway will remain in my memory as one of the most exhausting of the trip. We then enjoy the rest of the journey to Addis Ababa on the Adama-Ziway Express Highway, which runs parallel to the A7 and has been under construction by the China Railway Seventh Group since 2015. Perhaps because it is so new or because the tolls are too high, this four-lane motorway with its three-span concrete overpasses is practically car-free Figs. 14–15.

Day 130, 22 July – Addis Ababa

The Addis Ababa Institute of Technology (AAiT) is part of Addis Ababa University (AAU) and is considered the best technical university in the country. Since 2012, it has been divided into various schools and centres, one of which is the School of Civil and Environmental Engineering (SCEE). It emerged from the Department of Civil Engineering of the College of Engineering, which was founded in 1955. In 1961, this college became part of Haile Selassie I University, which was renamed Addis Ababa University in 1974. The SCEE website lists around 100 teaching staff – lecturers, assistant professors, associate professors, and full professors.

At the entrance to the civil engineers' building, very close to the National Museum, a marble plaque commemorates the laying of the foundation stone by German Federal President Heinrich Lübke on 24 October 1964. Professor Abebe Dinku points this out to me in German. This dignified, grey-haired gentleman with a thin moustache once gave my father a tour of his domain here. He received his doctorate from the University of Stuttgart in 1996 and is currently Vice-President of the AAU. Since 2007 he has also been involved in the Dialogue on Innovative Higher Education Strategies (DIES) programme of the German Academic Exchange Service (DAAD) as coordinator of various projects. The laboratories and test halls still contain the glass-fronted wooden display cases and models that were brought over from German-speaking countries back then. They are reminders of companies that have long since disappeared, such as the St. Egydyer Eisen- und Stahlindustriegesellschaft. To this day, there is still a strong connection with Germany. Some of the professors I meet during my visit speak German! Before German reunification, they had studied either in East or West Germany, with socialist or capitalist government funding, respectively. Later, in the summer of 2023, I invite a research assistant from Ethiopia to my department at the TU Berlin through an Elsa Neumann scholarship funded by the State of Berlin, and in the autumn of 2023, I meet Abebe again in Berlin. He will also contribute a text to the book. We discuss the possibilities for cooperation between our departments and whether he can co-supervise my doctoral students. Perhaps this will help to further develop good relations with Germany.

Day 131, 23 July – Addis Ababa 2

Buildings from the 1960s, both in the newly independent countries of Africa and in Ethiopia, clearly reflect the brutalist and structuralist tendencies typical of the time. As in Lusaka and Nairobi, Addis Ababa boasts impressive concrete buildings such as the National Bank from 1965, with its central light-flooded dome made of V-shaped concrete elements Figs. 16–17. The 1960s were a time of change and growth everywhere. Hexagons were part of the canon of forms at the time; indeed, they were fashionable, and they also made their way to Addis Ababa. The Filwoha Hotel and Baths, built in 1964 in the city centre, consist of a series of hexagonal structures, which in turn house hexagonal rooms for baths, showers, and changing rooms. The mosaic-tiled masonry walls of the Filwoha were covered with precast concrete elements, hexagonal truncated pyramids with skylights. This made it possible to build quickly and inexpensively. Lightweight and, therefore, resource-saving, durable, modular, and reusable precast concrete elements for roofs and ceilings are more relevant than ever in today's sustainability debate. The builders of yesteryear showed us how it's done, and we can learn from them. In my search for the designers of the Filwoha, I quickly find the architects, Israeli Zalman Enav and his Ethiopian partner Michael Tedros, who, with the support of Emperor Haile Selassie, helped shape the architecture of Ethiopia at that time. It takes me a little longer to find the name of the civil engineer. Ephraim Spira came to Addis Ababa in 1959 from

16

17

18

19

20

Figs. 16–17 The National Bank in Addis Ababa

Figs. 18–20 The Filwoha Baths in Addis Ababa

the Technion in Tel Aviv, the MIT of the Middle East [9]. He became the first dean of the aforementioned College of Engineering, which was later financed by Germany as part of Haile Selassie I University and headed by German dean Richard J. Bisanz from 1969 to 1974. Among the books Spira ordered for the University College were those by Buckminster Fuller and Pier Luigi Nervi, the masters of lightweight construction and concrete shells at the time. [9] Now I can contextualise the structure of the Filwoha Baths a little better Figs. 18–20.

In 2022, work is once again underway on a large bank in the city centre. With 52 floors and a height of 209 metres, the headquarters of the Commercial Bank of Ethiopia is the tallest building in East Africa. This skyscraper with a diamond-shaped floor plan, clad in a glass façade with striking vertical aluminium strips, was built by the China State Construction Engineering Corporation. European and American standards are used for Ethiopian construction, which are then adapted to the Ethiopian situation in terms of wind, earthquakes, and temperature loads. Dr. Abraham Assefa from AAiT is responsible for construction supervision, and Abebe Dinku asks him to take me to the construction site. He and one of the Chinese site managers proudly show me the elegantly furnished executive floor on one of the top floors, where the last pieces of furniture are still being assembled. I marvel at the fantastic panoramic view of the city stretching to the hilly horizon and recognise the hexagons of nearby Filwoha. In the control room on the ground floor, Chinese technicians sit intently in front of a wall of screens, adjusting the security systems. Another 20 metres below, in the foundations driven deep into the ground, with rooms full of emergency generators and a high-tech garage, I am impressed by the enormous round concrete pillars on which the building stands Figs. 21–23.

The design is the result of an international competition won in 2010 by the German architectural firm Henn together with the civil engineers from Buro Happold, the façade designers Priedemann, and the climate engineers from Transsolar. When I later tell Gunter Henn about my visit and ask him about his planning experiences, he is quite surprised because he has not heard anything from Addis Ababa since the winning competition entry, which his son Martin had supervised. In fact, he does not recognise

21

22

23

Figs. 21–23 The Commercial Bank of Ethiopia in Addis Ababa: façade, executive floor, and control room

24

25

Figs. 24–25 Lalibela: the interior of the Church of Biete Gabriel Rafael and the famous Church of Biete Giyorgis without any tourists

anything from the original design in the photos I show him. Another international competition in which even the winner came away empty-handed.

The four days I spend in this city are not enough to get me acquainted with its disturbing construction and traffic chaos. At least I get to visit Lucy. The skeleton of this early human of the genus Australopithecus afarensis, who already walked on two legs, is on display in a glass case at the National Museum. The 3.2-million-year-old lady, famously named by archaeologists after the psychedelic Beatles song *Lucy in the Sky with Diamonds*, proves that one of the cradles of humanity was here in the Horn of Africa.

Day 134, 26 July – Lalibela

In the Amhara region of northern Ethiopia, the Christian emperor Lalibela, a contemporary of the German emperor Barbarossa of the Staufer dynasty, ruled around 1200. According to tradition, he was commissioned by God to rebuild Jerusalem in Ethiopia. The legend says that pilgrims were to be given an alternative to the dangerous journey to the Holy Land. In this fairy-tale land of stone carved out of the rock, we walk from Adam's tomb to Golgotha. The Jordan River flows through the grounds, and to the right and left of it there are a dozen wondrous, very different rock churches. Many metres deep, some very narrow trenches have been carved into the rock to expose the church buildings or connect them. Their walls are covered with a thick layer of deep green moss, and below, water splashes over the shiny red rock floor. It is the height of the rainy season, and as I walk through these dripping trenches from one attraction to another, I feel like I am in another world. In contrast, we reach the Church of Merkurios in dry conditions via a long, pitch-dark tunnel, in which we feel our way forward, crouching down and holding onto the walls.

It is not only because of the rainy season that we are almost the only tourists at this World Heritage Site in July 2022. Most hotels are closed, and for a year now there has been no public electricity supply in Lalibela. There is hardly any running water and no cash machines. The war is close. Nevertheless, many hundreds of Ethiopian pilgrims are here visiting the appropriate church depending on their 'name day'. Today a service is being held in St Gabriel's Church. Before we enter, we leave our wet shoes at the entrance. Inside, it is dimly lit and we huddle with the faithful, soaked from waiting in the rain and wrapped in white cloths around the priests who are playing music, singing, and praying. Candles and incense envelop this scene in a thick heavenly mist. I have to get out Figs. 24–25.

After handing in our shoes again, we are able to study the other churches from the inside at our leisure. They are wonderful examples of subtractive manufacturing. The local red volcanic tuff was removed in such a way that free-standing monoliths surrounded by ditches or cave churches carved out of rock faces were created. The exposed rock formations were then hollowed out, probably from top to bottom, in such a way that church interiors of various shapes were created. With this type of construction, there is no room for error; the builders were at work here with precise measurements and impressive craftsmanship. Inside, structures were created that support a replica early Christian Basilica as well as simple, almost cave-like church interiors. In these churches, some of which have several storeys, we find free-standing round and square supports, and pilasters or pilaster strips protruding from the walls. They support beams, arches, flat ceilings, or barrel vaults. Or is this just an illusion? Which supports and ceilings really bear weight and which are only there for formal reasons? It would be interesting to analyse these monolithic structures as volumes and examine which of these forms are load-bearing. While the archaeology, history, architecture, and painting of these 12 churches and chapels of Lalibela – "four monolithic churches in the strict sense, the rest being tomb churches in all varieties of detachment from the mother rock" – is described in detail [10], their supporting structures seem to have been little studied to date. Abebe Dinku, my colleague from Addis Ababa, believes that we should definitely launch a research project on the supporting structures of the churches Figs. 26–29.

The stone 'monkey heads' that adorn the window corners of some facades in Lalibela are clearly ornamental. They are replicas of the wooden beams of Aksumite churches, protruding from the masonry walls, inserted perpendicular to the wall. There, together with alternating wooden longitudinal beams, they help to reinforce the brittle masonry against earthquake loads Figs. 30–31.

Since 2008, temporary roofs – membrane-covered space frames on slender steel supports – have been protecting some of the churches from rain and decay. Lalibela certainly needs this donation from the European Union. However, it is now almost impossible to photograph these churches without modern additions Fig. 32.

Day 137, 29 July – Gondar

The Italians, who were in the country between 1935 and 1941, wanted to administer northern Ethiopia from Gondar. Their brief occupation is commemorated by the Ethiopia Hotel in Gondar, which now serves only as a café, and the buildings I saw in Asmara, Eritrea. Here we find refuge from the desperate tour guides who, due to a lack of customers, won't leave the few foreigners on the street alone. After answering the introductory "where are you from" and repeatedly thanking them and informing them that I have already taken a guided tour of the city, I am bid farewell by a disappointed guide with a painful "You Hitler" Figs. 33–34.

Before that, we visited old Gondar, which was the capital of Ethiopia for around 200 years from 1636. The most important buildings from this period are located in the Royal Enclosure, foremost among them the four-storey palace of Emperor Fasilides, which curiously combines Ethiopian architecture with Indian Mughal turrets and Portuguese castle elements. The buildings show how interconnected the Ethiopian empire was in the seventeenth century. When I look at the buildings from the outside, I often see windows with arched tops, but on the inside I see horizontal wooden window lintels. Couldn't they make up their minds? Neither the guide nor his colleagues can help me here.

26

27

28

29

Figs. 26–29 The 'supporting structures' of the churches of Lalibela have not yet been examined in detail

30

31

32

Figs. 30–31 Wooden 'monkey heads' (31) at the church of Yemrehana Krestos (19 kilometres north-west of Lalibela) and stone ones in Lalibela at the church of Biete Amanuel (30)

Fig. 32 Canopy over the church of Biete Maryam, Lalibela

Perhaps the intention was to display beautiful arches on the outside, but have rectangular shutters on the inside? Some walls are supported by modern steel structures. I hear that Italian restorers seem to have caused some damage here by replacing dilapidated building ceilings with sturdy concrete slabs, which then overloaded the masonry of the walls. The masonry walls and towers of the emperors' palaces, baths, and stables are made of "basaltic rubble and lime mortar" [11] Figs. 36–37.

The 'Portuguese' bridges from the time of Emperor Fasilides, which still stand at the gates of the city, are also made of this masonry. It is raining so heavily that, despite my car's powerful engine, I can hardly make any headway on the muddy routes off the main road. But at least the Land Cruiser manages to reach the nearby Defecha Bridge over the Angereb River. This bridge, overgrown with greenery and supported by four circular arches with a stone archway on the side facing away from the city, is still used by pedestrians today and has certainly never been restored. I stand in awe before this endangered old lady, which has survived for almost 400 years and is certainly one of the first stone or masonry bridges in sub-Saharan Africa. I will have to postpone the search for the other bridges until my next trip. There are said to be around a dozen in the Gondar area [7, 12] Figs. 38–39.

34

35

Figs. 34–35 The Ethiopia Hotel on Gondar Piassa

36

37

38

39

Figs. 36–37 Palace of Emperor Fasilides. Why do the window openings have arches and wooden beams?

Figs. 38–39 Defecha Bridge near Gondar

A Brief history of Construction in Ethiopia
Prof. Dr. Abebe Dinku and Dr. Bikila Teklu,
AAiT, Addis Ababa University, Ethiopia, April 2024

'The Addis Ababa Institute of Technology (AAiT) at Addis Ababa University is the leading institute for engineering in Ethiopia. It has laid a solid foundation for the development of engineering education and research in Ethiopia. Its direct and indirect role in Ethiopia's economic development is immense. The institute has passed important milestones on its way to its current status.

In 1952, the College of Engineering of Addis Ababa was established on the premises of the technical school known as Tegbare-Id. The college was officially opened by His Imperial Majesty Haile Selassie I on 11 May 1955. Initially, the course lasted two years, with graduates receiving an intermediate diploma before going abroad for further studies. At the beginning, the number of students was quite small, but by 1954 it had grown to about 30, six of whom were sent abroad. Almost all of them earned their bachelor's degrees at renowned foreign universities and colleges. In the winter semester of 1955/1956, the university expanded the diploma programme to a four-year course with departments of civil engineering and industrial engineering. In 1958, the first bachelor's graduates received their diplomas from His Imperial Majesty Haile Selassie I. The entire programme was under the auspices of the Ministry of Education. The first dean was the American Robert Louis, followed by another American, Mr. Squires. The curriculum was divided into a two-year preparatory engineering course, completed with an intermediate diploma, and a three-year professional specialisation in civil engineering and industrial engineering.

In the winter semester of 1959/1960, Ethiopia and Israel signed an agreement whereby Israeli citizens would take over the administration of the university. Ephraim Spira held the office of dean. Other Israeli lecturers were appointed in various departments, particularly in the departments of civil engineering and electrical engineering. The State of Israel agreed to offer scholarships to graduates of the university to train them as future lecturers in Israel. In August 1961, two graduates from the civil engineering department and one graduate from the industrial engineering department were sent to Israel. In 1961, various universities merged to form Haile Selassie I University. These included the College of Engineering.

When the Derg military junta came to power in 1974, the name of the university was changed to its current name, Addis Ababa University. By 1973, a number of highly qualified Ethiopians were already employed at the university. Staff training received a major boost when agreements were reached with the government of the GDR that enabled Ethiopian teachers to study at universities there. Similar agreements were reached with the FRG. In later years, a number of faculty members were sent to study in the USA, the United Kingdom, Japan, etc.

As part of a bilateral technical assistance programme, the government of the Federal Republic of Germany kindly agreed to construct and hand over a new building for the future College of Engineering to the people of Ethiopia. The foundation stone was laid by the then Federal President of the Federal Republic of Germany, Dr. Heinrich Lübke. The university was to have fully equipped laboratories for civil engineering, electrical engineering, chemistry, and mechanical engineering. German technicians and lecturers were to run the facilities until sufficient Ethiopian personnel were available. The expansion of the College of Engineering led to the incorporation of the Building College in 1961, which had been founded in 1954 on the basis of a bilateral agreement between the Swedish and Ethiopian governments. In addition, architecture education began in 1963, promoted by a development and expansion programme. In 1971, the new building, constructed and equipped by the Federal Republic of Germany, was inaugurated by His Imperial Majesty Haile Selassie. With remarkable dedication, the various departments of the university have improved their range of courses over the years at both undergraduate and graduate level in order to compete internationally and contribute to the solution of specific engineering problems.

In 1955, two years after the founding of the College of Engineering, civil engineering was introduced as a four-year bachelor's degree programme. In 1959, the duration of the programme was extended to five years to make it easier for students to obtain the necessary qualifications for a career in civil engineering. The primary goal of the department was to train civil engineers who could adapt flexibly to the varying requirements of potential employers and become productive in the construction industry as quickly as possible after graduation. The Civil Engineering Department began offering a two-year full-time master's programme in structural engineering, hydraulic engineering, and geotechnical engineering in September 1979. The most important reason for introducing the master's programme was to train students to become qualified specialists who can tackle complex problems in the construction industry with little or no guidance from professors and supervise the designs of inexperienced engineers working in a project team – in short, to become highly qualified leaders. Such specialists play an important role in the successful implementation of national development programmes. The Department of Civil Engineering took another major step in this direction in 2005 with the introduction of a doctoral programme.

As part of a structural reform at AAiT in 2012, the Department of Civil Engineering was upgraded to the School of Civil & Environmental Engineering (SCEE). The aim of the SCEE is to equip its students with the skills they need to excel in technical decision-making and design, take on leadership positions in engineering firms, and undertake teaching in master's programmes and challenging research in engineering. The faculty has teaching and research laboratories for building materials and structures, road construction, hydraulic engineering, urban water management, hydroinformatics, and geotechnical engineering.

Over the years, civil engineering graduates have taken on leadership positions (including ministerial and prime ministerial posts) and managed major projects as representatives of building owners, engineering firms, or construction companies. Many have also pursued scientific and professional careers in Africa, Europe, and the US. From the outset, the SCEE has collaborated with a wide range of international and national partners. Overall, Germany's contribution to the advancement of civil engineering in Ethiopia is unprecedented in many respects.'

This essay is an excerpt from a book on the history of AAIT, which is currently being prepared.

EGYPT
LIBYA
Luxor
Aswan
Abu Simbel
Wadi Halfa
Kerma
Day 150, Dongola
Old Dongola
Port Sudan
Suakin
Day 146, Bayuda Desert
Kabushiya
Naqa
CHAD
SUDAN
Omdurman
Khartoum
Day 142+143, Khartoum
Kassala
ERITREA
Aksum
Gheralta
Mekelle
Al-Fashir
Nyala
Gonder
Lalibela
Bahir Dar
Addis Ababa
SOUTH SUDAN
0 km
250 km

Sudan
31 July–15 August 2022

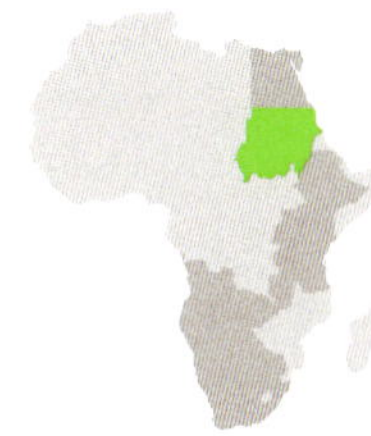

I'm back in Sudan after 45 years. I feel well in the heat and sandy deserts here at the transition between the Sahara and sub-Saharan Africa. I appreciate Sudanese Sufism with its dervishes and dances, a mystical and spiritual branch of Islam that I find quite tolerant, as well as the unobtrusive hospitality of the Sudanese people. I am amused by the fact that from the southern border near Gallabat to Wadi Halfa in the north, the same tin tea and coffee stand with a glass counter and the same strict but humorous saleswoman seems to be everywhere. In no time at all, she brews us a hot drink on a small charcoal stove, and her comments about us make the bystanders laugh. This quickly gets us talking – in sign language, because nothing works in rural Sudan without knowledge of Arabic. At least I have memorised the Arabic alphabet and can decipher road signs. That's why Khalid, a painter and tour guide from Khartoum, accompanies us during our two weeks in Sudan. He seems to know every detail of the country's millennia-old, incredibly rich history, whose stone legacies along the Nile he brings back to life with his stories.

Khalid is a calm and level-headed communicator who guides us through the regular roadblocks and saves us time because he knows when it's the police (drive slowly through and continue) and when it's the secret service (stop and politely show your papers). He also knows when and where we can travel without getting caught up in the regular demonstrations against the military regime. The situation in the country is 'unstable'. Fortunately, the military has ousted the previous ruler, Umar Hasan Ahmad al-Bashir, who only harmed and divided the country. Now it seems to want to stay in power and, as is also happening in Egypt, is establishing a state within a state, which is not good for the country. The population is resisting and going on strike or working to rule. This is evident because rubbish collection is no longer functioning, as reflected in the state of the country. Credit cards cannot be used to pay, and ATMs are not dispensing cash. The only way to get through the country is with new US dollar bills. At every petrol station, we have to hope that there is petrol available, but this is the case much more often than in Ethiopia.

Day 142, 3 August – Khartoum

In Khartoum, we stay at the centrally located Acropole Hotel, a charming classic hotel for travellers to Sudan. It has been run by a Greek family for over 50 years. Finally, a clean room with running water, electricity, and air conditioning. From here, it is less than two kilometres to the University of Khartoum, where I am giving a lecture on lightweight structures at the Civil Engineering Department of the Faculty of Engineering today. Around 15 full, associate, and assistant professors teach at the department, which was established in 1939. Dr. Amged Abdelatif, Assistant Professor of Civil Engineering, invited me and first introduces me to Professor Dafallah Al-Turabi, who was the first Sudanese dean of the faculty from 1963 to 1969. What an honour that this distinguished figure, dressed in traditional jellabiya and a white turban, took the trouble to come! An audience of over 150, at least a third of whom are women, sit in the large lecture hall with its red folding chairs and large screen Fig. 1. My conviction is that in the long term we will have less of an energy problem and more of a resource problem, and that resource-saving lightweight structures will, therefore, become increasingly important. This triggers an interesting discussion about solar energy and traditional construction methods among the audience in a country that is currently struggling with exploding diesel prices. The question and discussion session then continues among the professors over sandwiches and soft drinks. Should the Republic of Sudan, which lost three-quarters of its oil fields with the secession of South Sudan, not rely more on solar energy? Could the country at least return to traditional mud construction in the rain-poor north and import less expensive steel? The problems that need to be addressed here are the same as those throughout Africa: affordable housing and infrastructure. Only 7 per cent of the population in Sudan has access to water and sewage systems I am told, and according to UNICEF, only 53 per cent of rural households have access to clean drinking water sources within a 30-minute walk.

We visit the campus with its beautiful masonry buildings. The Building and Road Research Institute conducts research in structural concrete. However, it is well known that even simple concrete buildings cost at least US$300/m², partly due to high taxes on concrete and steel, making them unaffordable for most Sudanese. That is why research is also being conducted in the field of local materials and a reinterpretation of vernacular houses using mud bricks and rammed earth is being tested.

1

2

Fig. 1–2 Lecture at the University of Khartoum (photo: Amged Abdelatif) and Gordon Memorial College

The most famous building on campus is Gordon Memorial College, now the university's main library (see also introduction; Fig. 27). It commemorates Charles George Gordon, the Governor-General of the Egyptian province of Sudan, who defended the city for a long time during the Mahdi uprising. He was shot dead in 1885 during the ultimately victorious assault by the Mahdists. [1] The venerable building, inaugurated in 1902 – the British had recaptured Khartoum – with its shady arcades, features a mixture of Romanesque and Gothic arches with elements of Islamic architecture Fig. 2. The building material is locally produced bricks, which are arranged in three layers to form the load-bearing walls. These walls are reinforced with Nubian sandstone at their corners. [2]

I am particularly surprised by the Exam Hall, built in 1963, whose jagged façade of fired clay bricks evokes associations with the works of Ulrich Müther and Félix Candela even from a distance. And indeed, the spacious, column-free interior of the exam building, which can accommodate up to 500 students, is covered by four composite hyperbolic paraboloids. Here in Khartoum, which is almost devoid of trees, stands one of the rare wooden HP shells. Brown mahogany wood from South Sudan was used at the time for the straight beams that form these double-curved ruled surfaces and for the cladding. The design is by British architect Alick Potter, who founded and then headed the Department of Architecture in 1957. Together with wood and prefabrication expert Ezra Levin, he wanted his design to provide a modern, locally oriented response to the colonial architecture that had prevailed until then. [3] Ezra Levin was also an architect. In 1960, he designed the shiny silver laboratory and living quarters made of foam-insulated boat plywood for Edmund Hillary's Silver Hut expedition in the Himalayas. Who was the engineer behind the roof in Khartoum? At the Department of Architecture, my colleagues invite me in for tea and show me a wooden model of the roof structure that the students built. Lightweight shell structures – whether made of wood or other materials – are sustainable because they are resource efficient, so I am pleased that they are also covered in the course. Hyperbolic paraboloids are high-performance structures, and I wonder why civil engineering is not also interested in them Figs. 3–4.

3

4

Fig. 3–4 The examination building at the University of Khartoum with its hyperboloid roof

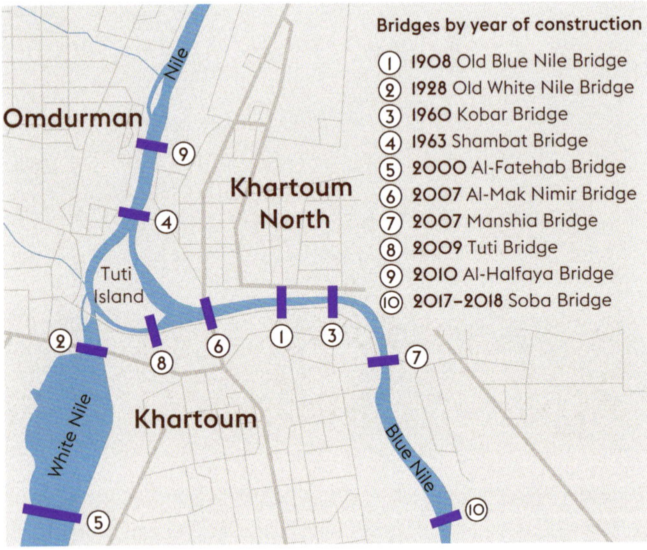

Fig. 5 The Nile bridges in Khartoum

Day 143, 4 August – Khartoum 2

The Ministry of Infrastructure and Transportation has invited me to give a lecture on bridge construction in Germany today. The German Foreign Office warns that photographing bridges in Sudan is strictly prohibited. Bridges are considered critical infrastructure. After the lecture, I remark how unfortunate it is that I cannot show the Nile bridges of Khartoum to my students in Germany. Immediately, a car with two ministry employees is made available to me. Off we go on a bridge tour, where I am allowed to photograph everything and inspect two more bridges. At over 40° Celsius, we are sweating, but I am happy. Khartoum boasts an interesting collection of ten large Nile bridges, built between 1908 and 2018. All kinds of bridge types, trusses, concrete box girders, cable-stayed bridges, and even a suspension bridge provide an overview here Fig. 5.

The second oldest structure in the Khartoum bridge family is the Old White Nile Bridge. This multi-span, 610-metre-long steel truss bridge, built in 1926, was initially used for rail traffic. To allow large ships to pass, it originally had a rotating bridge section approximately 93 metres in length. It was designed by the London engineering firm Fox, which had also designed the arch bridge at Victoria Falls in Zambia. [4] The trusses remind me very much of the Second Vivekananda Bridge over the Hooghly River in Kolkata. My companions, both civil engineers, believe that the Old White Nile Bridge was actually planned for India, but then ended up in Khartoum for unknown reasons. I cannot verify this. The fact is, however, that the bridge, which was widened in 1962 and is now used for road traffic, was in such poor condition at the turn of the millennium that the South African company Retro Construction Limited was commissioned to renovate this old-timer, and the German engineering firm Schüßler-Plan assessed and verified the restoration calculations. The sliding bearings under the truss girders were irreversibly rusted solid, and the girders thus formed frames with the piers, which led to large constraints and, mathematically, to overloading when temperatures changed. Because the bearings could not be replaced, demolition of the bridge was on the cards. However, the planners from Berlin were able to prove to the client that the bridge could still be used with local reinforcements and restrictions on heavy traffic. [5] Fifteen years after its reopening in 2007, the busy, almost 100-year-old vintage bridge is still in good, rust-free condition and is a fine example of durability, good adaptability, and successful bridge restoration Figs. 6–7.

The 210-metre-long Tuti suspension bridge was only inaugurated in 2009. A cable-stayed bridge would be more economical for such a length because the cables and abutments of the suspension bridge are expensive. It was certainly prestige that tipped the scales in favour of this bridge type in the centre of the city near a luxury hotel nicknamed 'Gaddafi's Egg'. The destination, Tuti Island in the Nile, was previously only accessible by ferry and offered mostly farmland, but now at least the city's residents have easy access to it. They place their plastic chairs on the sandy beach to cool their feet in the Nile water. Dorman Long Technology, whose predecessor had already built the Old White Nile Bridge, was responsible for the planning and execution here. The steel-concrete composite superstructure of the bridge is supported by two main suspension cables, each consisting of nineteen 70-millimetre-thick spiral cables. [6] We inspect the hangers that connect the suspension cables to the superstructure because my companions are concerned about their condition. I see no serious damage and draw their attention to the sand that has accumulated 10 centimetres high at the hanger anchors on the

6

7

8

9

Fig. 6–7 Old White Nile Bridge in Khartoum and Second Vivekananda Bridge in Kolkata, India

Fig. 8–9 Tuti Bridge with 'Gaddafi's Egg' in the background and detailed view of the hangers

superstructure. This should be swept away, as moisture can form here and cause corrosion Figs. 8–9.

Unfortunately, the senseless civil war in Sudan, which has been raging since April 2023, has also claimed a bridge victim in Khartoum. On 11 November 2023, part of the Shambat Bridge connecting Khartoum North with Omdurman was blown up. As always, the warring parties blame each other. [7] Now that this concrete box girder bridge is missing a section, traffic will certainly divert to the Old White Nile Bridge to the south. Hopefully it will be able to cope.

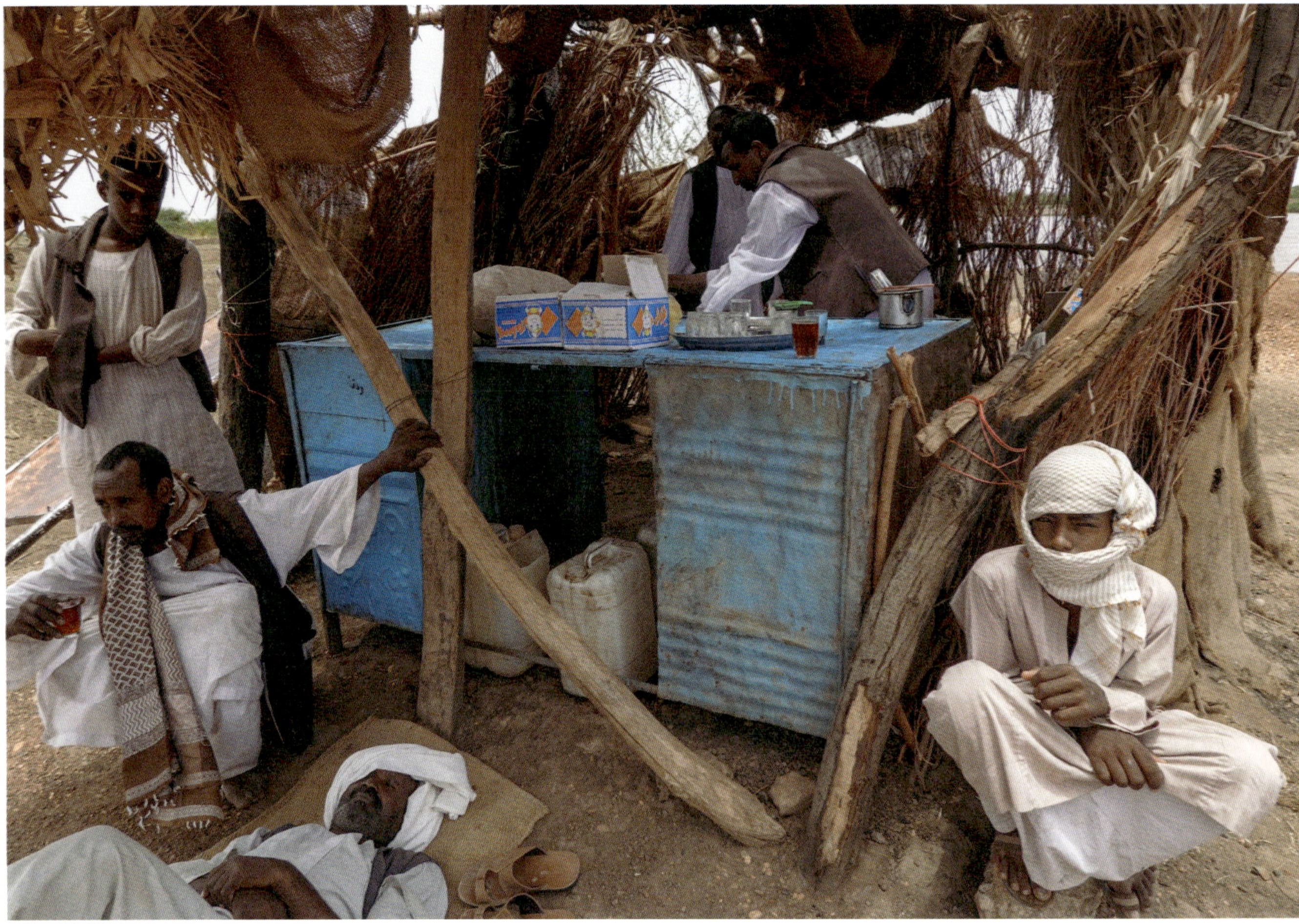

10

Fig. 10 Ferry station on the Nile near Mukrabiya

11

12

13

Fig. 11–13 The pyramids of Meroe, in the Bayuda Desert, and the Ottoman Well there

Day 146, 7 August – Bayuda Desert

About 30 kilometres north of the pyramids of Meroe Fig. 11, we turn left off the asphalt road to Atbara at Mukrabiya and, after driving for a kilometre past irrigated fields, arrive at the banks of the Nile. There, a donkey cart and an old Toyota pickup truck with five barrels of petrol are already waiting for the ferry. The thermometer reads over 45° Celsius. In the shade of a thatched roof, where tea is being prepared for us at a small counter, other travellers are dozing. We buy our tickets, sip our sweet drink, and doze off too Fig. 10. A little later, an old, somewhat rusty barge pushes its steel ramp onto the sandy banks of the Nile, and with a little momentum, all the vehicles safely make it onto the deck. The captain, relaxed and seated on a stool welded together from reinforcing bars in front of the steering wheel, manoeuvres the ferry across. We drive another two kilometres past palm trees and fields, then the Bayuda Desert opens up before us. Sand, stones, and a few dry bushes stretch to the horizon, but apart from a few tyre tracks, there is no road. Khalid guides us through this deserted area with the uncanny certainty of a sleepwalker Fig. 12.

After a few hours, we arrive at Ottoman Well, also known as Ababda Well. So, here in what seems to me to be the middle of nowhere, people do live after all. They come from far away with their camels and donkeys to this well to get water.

14

15

Figs. 14–15 Oasis with solar-powered water pump

Khalid goes ahead, talks to the head of the family, and presents one of our last tins of food as a gift. Then we are allowed to watch as the precious liquid is drawn from the depths of the well using a leather bag. Above the well opening, a roller is attached to an adventurous wooden structure, which a donkey uses to pull the rope away from the well until the dripping bag appears and can be grabbed. Two boys immediately pour the water into goat skins sewn together into bags, and the donkey is led back to the well. The procedure is repeated until all the water bags are full and the animals are watered. It happens quickly because the next family is already waiting under a small, shaded roof. It doesn't take much imagination to picture that things were exactly the same here in biblical times Figs. 13 and 22.

Less than 300 kilometres west in the same desert, a green garden passes by in the distance. A mirage? We turn around: the garden is real and inhabited. Khalid asks if we may come closer. We still have a 'Rei in the Tube' (the classic German travel detergent) that we give to the owner of the house, and we are invited to tea under green bushes. Clear water, very different from the murky well water from a few days ago, splashes into a small pond where a few ducks are swimming. How is this possible? We are given a tour of the desert estate, which consists of a small mud house and a stable. Solar panels are installed on the roof, producing enough electricity to power a suction pump. It draws a jet of water as thick as an arm from a depth of around 15 metres. It's that simple. The groundwater here is not deep, allowing the farmer to plant sesame in the sand and lead a very different life from that at the Ottoman Well Figs. 14–15.

Does the groundwater come from the huge but exhaustible Nubian sandstone aquifer system? No, I don't know how long groundwater will remain available in the Bayuda Desert, and I am aware that soil irrigated in this way can become salinated at some point. Nevertheless, from the ancient draw well to the pump that draws fresh water using solar energy, developmental steps such as the steam engine and the diesel-powered pump are simply skipped. The fact that this is possible makes me feel confident.

Day 150, 11 August – Dongola

In Nubia, northern Sudan, a variety of stone and masonry artefacts from a wide range of past cultures and ethnic groups have survived along the Nile between Khartoum and Wadi Halfa. From 2500 to 1500 BCE, the Kerma culture and the resulting Kingdom of Kerma existed here. Then came the Egyptians, who were replaced around 500 years later by the Kingdom of Kush with its capitals Napata and then Meroe. From 500 CE, Christianity spread throughout Nubia, but was later slowly supplanted by Islam until around 1500. After the Funji Sultanate (1504–1821), the Turkish-Egyptian Empire took power, and the suppression of the Mahdi uprising (1881–1899) was followed by British occupation until the country gained independence in 1956.

At Old Dongola, Sufi tombs called qubba stand like oversized beehives in the sandy terrain not far from the Nile Fig 16. These shell structures, measuring up to around ten metres in height and made of unfired mud bricks, which I would describe as rotational pointed arches, were built in the seventeenth century. [8] They are formed by rings of horizontally arranged stones lying on top of each other, whose radius becomes smaller and smaller towards the top. Such shells are called 'false' vaults because the compressive forces that have to be transferred downwards in the shell plane must be guided diagonally across the stone surfaces. The friction between the stones must be sufficient to prevent them from sliding outwards. False vaults are also found in the trulli made of natural stone in Apulia, Italy. The dome of the Church of Santa Maria del Fiore in Florence is, on the other hand, a real vault. The architect Brunelleschi inclined the masonry rings of this dome so that the shell forces press perpendicularly onto the stones and slipping is no longer possible. This allows greater forces to be transmitted. Nevertheless, the qubba are impressively slender, probably because the mud bricks are baked together. Their wall thickness is sometimes only the width of a brick, as I can see from the shoulder-high entrances and various collapsed areas where rain has dissolved the shells over time.

16

17

Figs. 16–17 Qubbas near Old Dongola and the Deffufa in Kerma

Two days later, we visit what is probably the oldest and largest mud brick building on the continent near Kerma. The Deffufa dates back to the second millennium BCE, measures 50 × 25 metres in plan, and reaches a height of 19 metres. It is unclear whether this eroded giant once served as a castle, temple, or tomb Fig. 17. The Kerma Cultural Centre museum, which opened in 2008 right next to the monument, provides an overview of the history of Nubia and the Deffufa, as well as the excavations in the surrounding area. It was designed by Sudanese architect Abdallah Sabbar and his firm Al Takamul Al Handasi in Dubai. [9] In keeping with fourteenth-century Nubian architecture, he constructed the roof from a series of barrel vaults, some oriented north-south and some east-west, at varying heights. The stiffening bulkheads at the ends of these roof elements have triangular openings, allowing the rooms below to be naturally ventilated in all directions. Inside, a few large fans provide additional ventilation. We stay in the pleasant cool environment much longer than in the almost unbearable heat outside in the excavation field, in the centre of which stands the Deffufa.

I even find an old bridge, or at least the remains of one, here in the sandy desert. In order to bypass the Nile, which is difficult to navigate in this area due to the second to fourth cataracts, the Egyptians and later the British built a railway line from Wadi Halfa to Dongola at the end of the nineteenth century. [10] It initially served as a supply line for the British troops under their commander Herbert Kitchener in the fight against the successors of the Mahdi Muhammad Ahmad. The line to Kerma was completed in 1897. On the right bank of the Nile, just a few hundred metres from the Ottoman Fort at the Jebel Wahaba burial ground, Khalid shows us the remains of Kitchener's Bridge. Only the three well-preserved pillars clad in local stone and the abutments of the four-span bridge over a wadi remain here – stripped of their superstructure – in the sand Fig. 18. The brick railway embankment is also well preserved and can be clearly seen on Google Maps (19.946390, 30.563049) all the way to Kerma. What an effort was made back then! As early as 1904, the builders had to watch as the line was decommissioned again. In the meantime, a more strategically sensible railway line had been built from Wadi Halfa through the Nubian Desert directly to Khartoum, thus bypassing the Nile cataracts 2 to 6. I travelled on this line in 1983. In Wadi Halfa, I visit the old railway station, where the journey to Khartoum began at that time. It stands there completely abandoned Fig. 19. Today the line is only used for occasional freight traffic. These military structures are not sustainable, and I hope that they will eventually be reused for civilians. Could these lines, used for peaceful purposes for the local population and Nile tourism, not experience a renaissance?

The last 100 kilometres to the border run through scorching hot desert, where only a few gold prospectors wander around with their metal detectors. One of these figures is walking near the road: a white cloth wrapped around his head and an ankle-length white robe protect the man from the sun. A heavy black work vest helps him carry the detector, which looks like a portable electric lawnmower. I would have loved to stop and ask if the business is profitable, but there is a holster hanging from our gold prospector's belt, and we don't want to find out what's inside it. We also drive past the eerie and deserted-looking camps of these people with their living containers, heavy machinery, and gold mining equipment without stopping. There is no greenery, no café, and no petrol station to invite us to linger here Fig. 20.

Later, hundreds of cattle carcasses line the roadside. Sudan supplies the slaughterhouses of Aswan in Egypt, and before the border, all animals that did not survive the journey from the south are thrown from the transporters Fig. 21. Doesn't that bother anyone?

18

19

20

21

Figs. 18–19 The pillars of Kitchener's Bridge at the Third Nile Cataract and the railway station at Wadi Halfa

Fig. 20–21 Gold prospectors in the Nubian Desert and cattle carcasses on the roadside outside Wadi Halfa

Overview of Construction in Sudan
Prof. Dr. Amged Osman Abdelatif,
University of Khartoum, Sudan, May 2024

'The history of civil engineering education in Sudan began in 1939 at the University of Khartoum. Since then, the field has experienced remarkable growth; there are now more than 50 degree programmes offering civil engineering studies at diploma, bachelor's, master's, and doctoral levels.

Sudan is blessed in geographical terms: Khartoum is located at the confluence of the White and Blue Nile, which form the majestic Nile River. The Nile then joins the Atbara River and flows north through desert landscapes and past more than 200 pyramids dating from 700 BCE to 300 CE. These historic structures are testament to the inventive spirit and engineering skills of our ancestors. The fertile banks of the Nile are home to numerous agricultural initiatives, the largest of which is the Gezira Project, covering an area of over two million hectares. The construction of the Sannar Dam in 1924 proved to be crucial, as it enabled the irrigation of the Gezira Plain and the generation of electricity. Several other dams were built in the following years, demonstrating Sudan's commitment to using its natural resources for the country's development.

The towns along the Nile are connected by numerous bridges, each of which is a testament to Sudan's ever-growing technical capabilities – from the historic truss bridges in Khartoum, which echo the past, to the towering elegance of the Tuti suspension bridge, a modern marvel that connects communities, people, and their dreams.

Although challenges remain, such as an inadequate road and rail network and the harsh climatic conditions on the Red Sea coast, the construction industry in Sudan is developing steadily and continuously.

Traditional Sudanese houses are usually single-storey and made of readily available local materials such as mud, stones, wood, fired mud bricks, and cement blocks, with a variety of roof structures. In contrast, reinforced concrete buildings and industrial steel structures are common in urban areas, reflecting the development of construction techniques and materials. In the major cities, the architectural legacy of British colonial rule, which lasted from 1899 to 1956, is still visible in the cityscape. The buildings constructed during that period, with their distinctive colonial style, continue to house government agencies, universities, and schools.

In recent years, the number of engineering firms and local construction companies has risen sharply. This has breathed new life into Sudan's architectural visions. International cooperation has brought new perspectives and expertise, further enriching the country's construction industry. The Sudanese Engineering Society and the Engineering Council play a crucial role in the further development of the industry, as they are responsible for accrediting civil engineers and ensuring the highest professional standards in this field. Despite this progress, the challenges that lie ahead are considerable. There is a huge demand for qualified civil engineers – engineers who will plan and build the important infrastructure of the future and advance the nation on its path to achieving its development goals. Their legacy is based on tradition, innovation, and an unwavering commitment to a better future for Sudan.'

Fig. 22
At the Ottoman Well
in the Bayuda Desert

22

Day 176, Alexandria
Port Said
Mahalla al-Kubra
al-Mansura
Tanta
Zagazig
Shubra El-Kheima
Ismailia
Cairo
Day 166, Cairo
Gizeh
al-Chusus
Suez
Fayyum
LIBYA
Day 160, Western Desert
Asyut
Luxor
EGYPT
Day 156, Aswan
Day 154, Abu Simbel
SUDAN
Kerma
Old Dongola
0 km
250 km

Egypt
15 August–6 September 2022

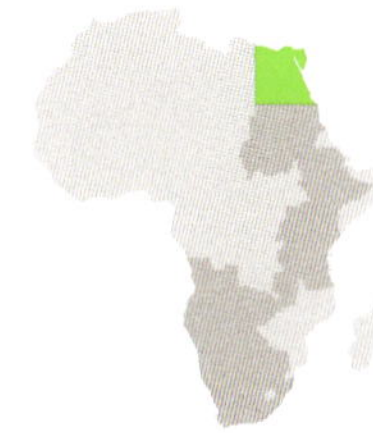

When asked whether or not he feels African, an Egyptian colleague gives me a three-part answer: "Geographically yes, historically no, and economically yes again because we want to be part of the booming Africa of the future." Indeed, Egypt differs from sub-Saharan Africa. As in Sudan, Islam and the Arabic language dominate here. It was and is directly connected to Asia and Europe thanks to the Mediterranean Sea. These influences, which have existed for thousands of years, have shaped the country differently from those countries of the south. At the same time, the Nile, with its sources in Kenya and Ethiopia, is a southern link for culture and trade that no other Mediterranean country in Africa has.

Along the main roads, I see advertising banners, as large as the biggest German election posters, for luxury apartments on the Mediterranean coast (near El Alamein) or on the Red Sea (near Ain Suchna). The advertised living spaces could also be found in the gated communities of Dubai or Kuwait; wealthy Egyptians from the big cities appreciate this internationally interchangeable chic with private access to the sea as a summer retreat. Non-Egyptian beach tourists are more drawn to Sharm El-Sheikh and Hurghada. From there, they take organised trips to Luxor and Cairo. There are virtually no tourists travelling in their own cars because the temporary importation of a private vehicle involves an incredible amount of bureaucracy and because car ferries do not currently serve the country.

The current construction boom triggered by the military is striking. Within a very short period of time, 7,000 kilometres of motorway have been built. Everywhere you look, there are comfortable, new, multi-lane expressways with hardly any traffic, which were obviously built in a great hurry and without extensive feasibility studies. One consequence is that in many places there is a lack of junctions, turning points, and, above all, crossings. The rail network is also to be expanded to an almost unimaginable extent. Around 2,000 kilometres of railway are to bring the country's traffic onto the rails and, if everything is actually implemented, make travelling much more comfortable. The trains come from Siemens [1], but everything else is 'Egyptian built' Figs. 1–2.

What is not clear to the observer, and for which even the engineers I interviewed have no convincing answer, is the question of the economic viability of these mega-projects and the speed with which they are being carried out. The construction industry is completely overheated with the military as its sole client, and engineering firms have to deliver their design projects in a rush and for ridiculously low fees. This can only be at the expense of quality and could leave everyone involved in a slump after the flash in the pan. And where does the money for all the major projects I have seen come from? According to a 2022 study by the German Institute for International and Security Affairs, Cairo's foreign debt has more than tripled since the military took power in 2013. [2]

1

Fig. 1 On one of Egypt's many new motorways

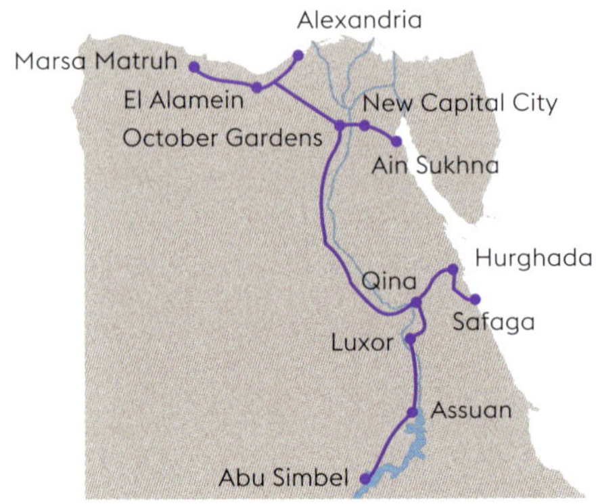

2

Fig. 2 The dark blue lines show 2,000 kilometres of rail network for high-speed trains to be built in collaboration with Siemens [1]

Day 154, 15 August – Abu Simbel

May I report on another border crossing? At the Egyptian border, our car is inspected by customs, drug enforcement officers, the police, the military, and the bomb squad. I need an Egyptian driving licence, Egyptian insurance, and Egyptian number plates. So, it continues. And because we had heard beforehand that the procedure could take several days without help, we enlisted the services of an agent in good time and sent him all the information and papers in advance. It cost us 1,000 euros, but with his support on the ground, we manage to cross our last intra-African border in just four hours.

A ferry takes us across Lake Nasser directly to Abu Simbel. I've been to Egypt half a dozen times, but just like the World Heritage Sites in the Valley of the Kings, Luxor, and Karnak, you can't just drive past here. We drink our first beer in over two weeks and get used to the first tourist groups in almost two months.

The World Saves Abu Simbel is the title of a book published in 1968 [3] that was on my parents' bookshelf and that I picked up again and again. In it, Egyptologist Christiane Desroches-Noblecourt and journalist and photographer Georg Gerster report on a fascinating feat of engineering in the Nubian Desert: The magnificent structures of Abu Simbel, dating from the thirteenth century BCE, the time of Ramses II, had to be relocated as the water rising behind the new Aswan Dam was about to make them disappear forever. At that time, 50 nations joined forces under the leadership of UNESCO to carry out the relocation operation and to rescue Abu Simbel. For the two largest structures at risk, the two rock temples of Abu Simbel – famous for the four colossal royal statues at the great temple – it was almost too late when in November 1963 an international bidding group, including Hochtief from Essen, Germany, agreed to move them to dry land after all. What followed was a spectacular operation in which an international community of archaeologists, engineers, technicians, and workers dismantled the rock temples under great time pressure and rebuilt them around 65 metres higher and 180 metres inland. To this end, the temples and parts of the surrounding sandstone mountain into which they were carved were sawn with the utmost precision into around 20,000 blocks weighing a maximum of 30 tonnes. In a "feverish rush", according to the book, the team first built a 372-metre-long and 25-metre-high dam so that they could continue working in dry conditions. The sawing and relocation operation was more successful than predicted, but the rebuilt temples with their mortar-plastered saw cuts were no longer able to support the mountain above them. That is why protective reinforced concrete domes, each composed of a cylindrical shell and a quarter sphere, span over each temple ever since. The metre-thick dome above the large temple has a span of 59 metres, a clear height of 19 metres, and it has to support up to 9 metres of rock above it. To reduce the effects of temperature, shrinkage, and creep stresses, the construction companies concreted the thick shell in 300 concreting sections and laid 4,500 metres of cooling tubes in the concrete to control the setting temperatures after concreting with cooled water. The work was completed after four years. YouTube makes it possible to relive this impressive engineering feat, which took place half a century ago Figs. 3–4. [4]

Today, I can no longer see any traces of all this construction work. I feel as if I am standing in front of a monolith that has always been here. Here, engineering work has helped to mitigate the damage caused by other engineering work. The 5,000-square-kilometre Nasser Reservoir helps to supply Egypt with hydroelectric power. But it also flooded parts of Nubian history because in the 1960s only the most important treasures were saved. I don't believe there is a universal solution to such dilemmas; we have to laboriously discuss them anew for each project.

3

Fig. 3 The colossal statues of Abu Simbel

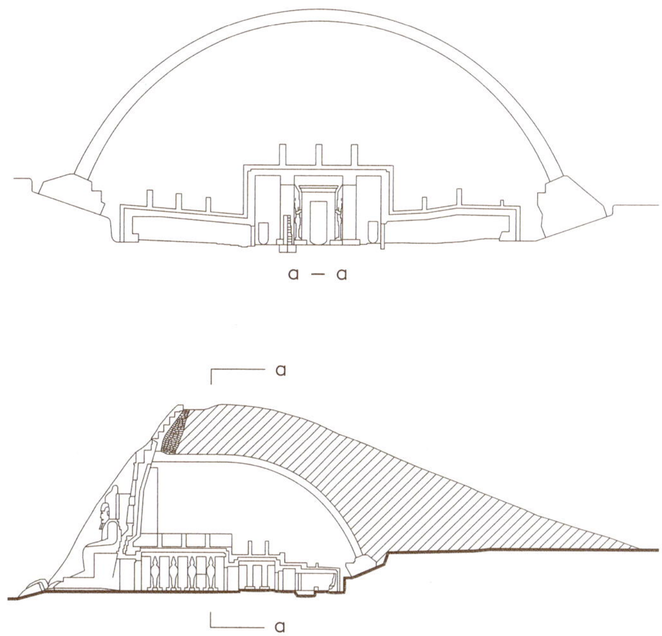

4

Fig. 4 Sketch of the reinforced concrete dome. Cross-section through the cylindrical shell and longitudinal section with stone deposit on top (by S.Thurath, sbp)

Day 156, 17 August – Aswan

Multi-storey concrete frame buildings, infilled with masonry and, if there is money left over, plastered, dominate the streetscape of the modern part of the city of Aswan. On the other hand, in the Nubian village of Aswan on the other, western side of the Nile, we find classic Nubian buildings, whose appearance is characterised by cylindrical shells and domes made of masonry Figs. 5–6. Fired bricks are used today for their beautiful roof structures, which are mainly subject to compressive stress and can, therefore, be constructed without steel reinforcement. The construction of such lightweight structures has a long tradition. Long before the Romans, such barrels and domes could be erected quickly and without the use of formwork in ancient Upper Egypt. In the arid south of the country, unfired bricks made from clay and sand washed up by the Nile are also used, making this construction method particularly sustainable and inexpensive. In the 1940s, Egyptian architect Hassan Fathy rediscovered this construction method while searching for architecture typical of the country. In his book *Architecture for the Poor* [5], he first states: "yet in modern Egypt there is no indigenous style. The signature is missing; the houses of rich and poor alike are without character, without an Egyptian accent." He then describes in detail his ideal of architecture, down to the exact composition of sun-dried mud bricks and the construction of vaults and cylindrical shells of various spans. He is concerned with 'rural reconstruction' and addresses issues of urban planning and the needs of the village community as well as climate issues, optimal orientation, and design of buildings for ventilation and shading. First published in 1969, the book is a primer on modern sustainable construction. Hassan Fathy's best-known project is the New Gourna settlement near Luxor. From 1945 to 1948, he was able to partially implement the ideals later summarised in his book. This resulted in numerous residential buildings, a market, a school, and a mosque. El Gourna was a resettlement project. The villagers who were to be resettled from Old Gourna to New Gourna eked out a meagre existence by plundering the burial grounds beneath their former village. The resettlement would have robbed them of this meagre livelihood, which mainly benefited others. Ultimately, the project failed due to resistance from these profiteers. Nevertheless, 60 per cent of what was built in New Gourna at the time has survived all these years, and the World Monuments Fund honours Hassan Fathy's work as an early experiment in sustainable construction. [6]

Just beyond Aswan, we cross the Nile on the first cable-stayed bridge since Khartoum. The 250-metre-long New Aswan City Bridge was designed by Arab Consulting Engineers (ACE) in collaboration with Rendel and built by the General Nile Company for Roads & Bridges. The two pylons of the simple and elegant cable-stayed bridge with a central cable plane support a torsion-resistant, 24-metre-wide concrete box girder with four lanes Fig. 7.

5

6

7

Figs. 5–7 Masonry domes in the Nubian village of Aswan and the new Aswan City Bridge with stay cables in one plane

Day 160, 21 August – Western Desert

Mahmoud Ezzeldin, an expert in desert driving, accompanies us to the Western Desert in his vintage 1975 Toyota Land Cruiser. With two cars, we can venture deep into this sandy wasteland. We lower the tyre pressure to 1 bar to maximise the car's contact area. Nevertheless, we keep getting stuck. We dig away the red-hot sand in front of the sunken tyres and place our aluminium sand plates underneath. Three of us push, and one carefully drives the car out of the rut in first gear with the four-wheel drive engaged and the rough aluminium under the wheels. This is how we slowly reach our destination, the White Desert Protected Area, which is known for its bizarre chalk sculptures shaped by the sandy desert wind. Under a wonderfully clear starry sky, we spend the night in the deathly silence of this desert, but not before searching the immediate vicinity of the cars for scorpion and snake tracks Figs. 8–11.

8

In the Bahariyya Oasis, about 60 kilometres to the north-east, I see houses and walls that appear to be built from the white desert chalk. I learn that the beautiful white stones are considered inexpensive and waterproof, but poor at insulating heat. Before the mortar is applied to the stones, the chalk dust must be thoroughly washed off, which is often forgotten and leads to poor bonding. The stones used to be delivered in sizes of 40 × 20 × 20 centimetres, but today they are 20 × 12 × 12 centimetres in size. I learn that they are not quarried here, but in El-Minya, a city 200 kilometres to the east on the Nile. There are extensive near-surface deposits of the purest white limestone, consisting of 99.6 per cent calcium carbonate. [7] The stones there are cut from the rock by large, electrically powered stone saws on wheels. On the website of *The Guardian* newspaper, I find pictures of workers enveloped in thick white clouds of dust, working in life-threatening conditions near noisy saw blades. [8] For all these reasons, red fired hollow bricks, which are manufactured in local factories, are preferred in Bahariyya.

9

10

11

Figs. 8–11 Chalk sculptures (8–9) and dune drive in the Western Desert (10–11)

Day 166, 26 August – Cairo

From a height of around 50 metres, I look out over the labyrinthine old town of Cairo with its many minarets and roofs adorned with a wondrous hodgepodge of satellite dishes, water tanks, bed frames, and makeshift extensions. This view is offered by the minaret of the Ibn Tulun Mosque, built of red fired brick, with its curious combination of a square base and a round tower with an external spiral staircase Fig. 12, probably dating from the thirteenth century. Completed in 1961, the Cairo Tower on the Nile island of Gezira stands at a height of 187 metres Fig. 13. The tallest building in Africa at the time, with its striking reinforced concrete shell perforated with diamond-shaped recesses, was designed by Egyptian architect, engineer, and contractor Naoum Shebib. This exceptional talent, who unlike Félix Candela or Eugène Freyssinet is hardly known in my part of the world, built Cairo's first high-rise buildings and left behind long-span concrete shell roofs. [9] His tower was built at almost the same time as Fritz Leonhardt's 216-metre-high television tower in Stuttgart, Germany, the world's first reinforced concrete tower. In the evening, my gaze wanders from the viewing platform over the majestically wide Nile, adorned with illuminated boats and slender bridges. The sea of houses in this city of 20 million stretches to the horizon. There, the triangular shapes of the pyramids stand out in the reddish haze Figs. 14–15.

We are also adding a new museum to the itinerary. Joining the Egyptian Museum – the grand old lady of museums on Tahrir Square (exhibition space: 10,000 m²), which opened in 1902 and features charming, old-fashioned wood-and-glass display cases and typewritten labels – is the National Museum of Egyptian Civilization (exhibition space: 23,000 m²), which opened in 2021. In addition to 50,000 exhibits, 22 mummies of famous Egyptian kings have also found their temporary final resting place there. For the design, Egyptian architect El Ghazzali Kosseiba – who had won the competition for the museum in 1983 – teamed up with Japanese architect Arata Isozaki. As is so often the case, nothing is known about the engineer. We cannot yet visit the large Egyptian Museum in Giza near the pyramids (exhibition space: 40,000 m²), whose construction began as early as 2002. It is scheduled to finally open in 2024.

It is also possible to admire a new cable-stayed bridge in Cairo. Opened in 2019, the Tahya Misr cable-stayed bridge, also known as the Rod El Farag Bridge, with its 300-metre span and 67-metre-wide superstructure, is considered the widest of its kind. [10] The long-established engineering firm ACE (Moharram and Bakhoum) from Cairo designed the bridge. I have known Mourad Bakhoum, now one of the directors of ACE, for many years and we meet regularly at events organised by the fib (Fédération internationale du béton) and the IABSE (International Association for Bridge and Structural Engineering). In his brief overview of the history of bridge construction in Egypt, he illustrates its long tradition and notes that the Egyptians only began building bridges over the Nile in the late nineteenth century. [11] His father, Michel Bakhoum (1913–1981), founded ACE in 1950 together with his partner Ahmed Moharram and is for Egypt what famous engineer Fritz Leonhardt is for Germany. The two great engineers lived at around the same time, influenced generations of engineers at the universities of their respective countries, and left their mark on building and bridge construction. One of these is the 6th of October Bridge, a concrete box girder bridge that crosses the Nile very close to the Egyptian Museum and forms part of a 22-kilometre-long traffic artery through the city. With his 1,400-page book *Structural Mechanics* [12], Michel Bakhoum has left behind a standard work of Egyptian engineering knowledge, and I am delighted when Mourad gives me one of the few remaining copies.

The hot Egyptian summer is a lecture-free period. Cairo University and Ain Shams University, considered the best state universities for civil engineers, are deserted during these months. Professor Salah El-Metwally from the equally important Mansoura University helps me and organises two lectures for me at the Housing and Building National Research Centre (HBRC) in Cairo Fig. 16. Fortunately, the lecture hall, equipped with the latest technology and a large screen, fills up, and so I see Fathy Saad from Ain Shams University again. He did his doctorate under my father and worked at schlaich bergermann partner for many years. He has since successfully established his own firm and was the engineer in charge of the construction company for the Rod El Farag Bridge. My colleagues here are convinced that the practical experience of Egyptian professors benefits the students. Unfortunately, this is also a necessity because

12

13

14

15

Figs. 12–15 The minaret of the Ibn Tulun Mosque (12) and the Cairo Tower (13). View from the Cairo Tower of the Nile bridges in Cairo (14) and the pyramids (15)

16

Fig. 16 Test hall of the Housing and Building National Research Centre (HBRC) in Cairo

Egyptian professors at state universities earn very little. A professorship is more of an honorary position with a prestigious title, but without a side job at an engineering firm, it is difficult to make ends meet.

Egyptian civil engineers are proud to be able to carry out all planning tasks themselves and in accordance with their own national standards. However, just like us, they lament the deterioration in training and craftsmanship, which often leads to structural elements being oversized for safety reasons. This 'more is better' approach is not sustainable in terms of minimising materials and is certainly one of the global construction problems that we all need to address.

Day 176, 5 September – Alexandria

By the sea, women in djellabas sit on white and blue plastic chairs and let the water wash over their feet while children romp around them. Ice cream vendors with hawkers' trays pass by. On the other side of the beach promenade, hotels from the 1930s line up next to each other. It is evening in Egypt's second largest city. We have reached the end of our tour Fig. 17.

We enjoy the feeling of having made it from Cape to Cairo. The fact that we are still together after almost six months in the car is not something to be taken for granted. We are saturated. But we don't want to miss out on the Bibliotheca Alexandrina, designed by Snøhetta architects in collaboration with the engineers at Hamza Associates. It won the IABSE Outstanding Structure Award in 2003. The stepped reading room, the world's largest with 2,000 seats, with its slender supports bearing the huge sloping roof, makes me linger in awe Fig. 18. In the adjoining museum, works by the Egyptian painter Hassan Suleiman, whom I met many years ago, are on display. What a coincidence! Four of his seagull paintings adorn my office in Berlin Fig. 19.

Alexandria and the Mediterranean herald Europe, but we cannot leave Egypt by land because we would not be able to pass through Libya or Syria. During the Arab Spring, the last ferry connection was also discontinued. We, therefore, ship our car from Alexandria to Derince in Turkey on a RoRo freighter and fly after it on 7 September.

17

18

Figs. 17–18 The beach promenade in Alexandria and the Bibliotheca Alexandrina

Fig. 19 Seagull paintings by Egyptian painter Hassan Suleiman

Civil Engineering in Egypt: Education, Research, and Practice

Prof. Dr. Salah El-Metwally,
Mansoura University, Egypt, June 2023

'In Egypt, civil engineering is a subject that leads to a bachelor's degree (BSc). At university level and after the bachelor's degree, it is divided into three specialisations: structural engineering, civil engineering, and irrigation and hydraulic engineering. These offer various graduate degrees: diploma, Master of Science (MSc), and Doctorate (PhD).

The bachelor's degree in civil engineering lasts five years. It is divided into one year of general scientific foundation courses and four years of civil engineering. During these four years, students take courses in materials technology, structural mechanics and structural analysis, concrete structure design, steel structure design, soil mechanics and foundation engineering, construction project management, surveying, sanitary engineering, road and airport construction, traffic engineering, hydraulics, urban water management, and port construction. The courses on concrete and steel structures cover the fundamentals of construction, buildings, large halls, hydraulic structures, prestressed concrete, earthquakes, traditional steel bridges, traditional concrete bridges, and special topics. In the final year of civil engineering studies, students must complete a final project in one of the aforementioned areas, working largely independently but under the supervision of faculty members.

For graduate degrees, students must pass a certain number of courses with a minimum grade. The diploma programme includes a final project. For the master's and PhD programmes, a minimum grade point average and passing an entrance examination are required. A thesis, supervised by at least two faculty members, is a demanding component of both degrees. An aptitude test must be taken for the PhD programme. In both programmes, the examination board for the defence of the thesis consists of at least one external faculty member, the main supervisor of the thesis, and a faculty member who is not one of the supervisors.

The courses are usually lectures with tutorials (for the basic courses), supplemented by practical exercises, assignments and tests, as well as intermediate and final examinations. Egyptian textbooks, lecture notes, and reference works, as well as international literature, are available for the courses in the bachelor's and master's programmes. In the design courses, the teaching material refers to Egyptian guidelines, such as the EPC 203 standard for the design and construction of concrete structures, the EPC 204 standard for the design and construction of buildings, the ECP 205 standard for the design and construction of steel structures, and the Egyptian regulations for the planning, design, and construction of bridges and elevated roads.

There are two professional associations for civil engineers: the Egyptian Syndicate of Engineers and the Egyptian Society of Engineers. The former deals with all legal and certification issues, while the latter is dedicated to the promotion of science and practice.'

OPEN

3 Considerations

Fig. 1 'Open House' (2017) in Cape Town, South Africa

Considerations

The following section contains seven short texts on aspects of construction in Africa that have been on my mind since the trip. I regard this series of thoughts as a basis for further discussion and for illuminating the topics from other angles. Decisions must be made soon. "There is no doubt that the future is bright for Africa," emphasised Akinwumi Adesina, President of the African Development Bank, in a speech marking the 40th anniversary of the British daily newspaper *The Guardian* at the end of 2023, referring to the growth figures. Here, too, the figures speak for themselves and demonstrate the potential for construction in Africa. Most of the data graphics are taken from the values in the table at the end of this book.

1. Universities – Skipping a Class
2. Solar Energy – Africa's Trump Card
3. Concrete – A Natural Building Material in Abundance
4. Bridge Building – A Wonderful Challenge
5. Roads and Railways – Not Yet Connected
6. China – Bringing it into the Team
7. Residential Construction – It's High Time

Universities – Skipping a Class

Institute or faculty of civil engineering, chair or department for conceptual and structural design? The academic designations in civil engineering are as varied in Africa as they are in Germany. When visiting universities, you have to navigate a complex maze of academic institutions, ranks, and degrees, which in the countries I have travelled to tend to be similar to the English-speaking, 'Anglo-Saxon' world.

In the US and Great Britain, and indeed in Japan, a strict distinction is made between civil and structural engineers. The former plan infrastructure projects such as roads, port facilities, tunnels, and bridges, while the latter are responsible for the supporting structures of buildings in collaboration with architects. The German term 'Bauingenieur' covers both civil and structural engineering. In Germany, as in Switzerland and Austria, engineers are trained as generalists and, after completing their studies, can work anywhere in the world and in many different fields. This also seems to be the case in Africa. The term 'civil engineer' is used in the same way as the German 'Bauingenieur'.

At the universities I visited Fig. 1, there is usually a 'college' , a 'school', or a 'faculty of engineering', divided into 'departments' for different engineering disciplines, such as a 'department of civil engineering'. The faculty is headed by a dean, and the respective department is under the leadership of a head of department (HOD) or chairman.

The academic degrees awarded in civil engineering are generally Bachelor (BSc), Master (MSc), and Doctor of Philosophy (PhD), which correspond to our German doctoral degree (Dr.). A diploma is almost exclusively awarded for educational qualifications below or outside of university degrees, such as for the completion of vocational training. The German system of dual apprenticeship training does not exist in many countries. Often people learn their profession through learning by doing on the construction site. In Namibia, however, craftsmen are trained at Vocational Training Centres (VTC).

The bachelor's degree in civil engineering (undergraduate) that qualifies graduates for a profession usually takes four to five years. In countries such as Zambia, Tanzania, and Uganda, a so-called A-level year is required beforehand where candidates are prepared for their studies. The focus of the teaching reflects the needs of infrastructure construction and the expertise of the teaching staff. In addition to basic subjects such as mathematics, mechanics, and statics, almost all universities teach the classic topics of road construction (paved and unpaved roads), geotechnical engineering and mining, structural hydraulic engineering, and urban water management.

Master's programmes (postgraduate), which are by no means offered by all universities, are designed to last two years. Compared to German master's programmes in civil engineering, they are heavily research-oriented. Lectures often make up less than a third of the master's programme. These programmes are rare because of a lack of facilities, money for expensive research activities, and experienced teaching staff. Where offered, the master's programme costs between 2,000 and 3,500 euros per semester depending on the university and is usually completed alongside work in evening classes. These 'evening students' do not have much time for research, which is why completing the master's thesis (dissertation) often takes significantly longer than two years. In contrast, German master's programmes, at least for civil engineers at the TU Berlin, are dominated by in-depth lectures, seminars, and practical exercises. The master's thesis, which is usually research-based, accounts for only 25 per cent of the required credits.

Teaching is carried out by lecturers, senior lecturers, associate professors, sometimes also assistant professors and, finally, at the top of the hierarchy, full professors. Not every civil engineering programme offers the latter. At some universities, all teaching is done by lecturers with master's degrees, who have to give up to 24 hours of lectures per week. Anyone who has to handle preparation, follow-up work, and exams in addition to so many hours of teaching will have little energy left for research.

While professors in South Africa can live well on their salaries, in Egypt a professorship at a state university, especially in civil engineering, is highly regarded but so poorly paid that virtually all Egyptian professors have to teach at a better-paying but less prestigious private university or work in an engineering firm. I am convinced that practical experience should be a prerequisite for professors in civil engineering. I know that authentic reports from the construction site and interesting examples from engineering practice increase students' motivation and understanding. However, financial hardship should not be the reason for taking on a second job in a firm or moonlighting at a second university.

Detailed general information on the educational landscape of many African countries can be found at the German Academic Exchange Service (DAAD). [1] You can even obtain an individual analysis of the education system in Egypt, Ethiopia, Kenya, and South Africa. [2] Germany began working with Addis Ababa University in Ethiopia in the 1960s, establishing the Institute of Civil Engineering there. In 2015, the German Society for International Cooperation (GIZ) donated the 'German Wing' to the University of Namibia (UNAM) in Ongwediva, where civil engineers teach and conduct research. Initially, it also seconded teaching as well as specialist staff from Germany and trained local teachers.

German universities of applied sciences and universities cooperate with African universities within the framework of exchange programmes, such as the University of the Federal Armed Forces in Munich with Stellenbosch University [3], or even offer a double degree in civil engineering, such as the FH Aachen together with the Namibia University of Science and Technology (NUST) in Windhoek [4]. Starting in 2012, the Technische Universität Berlin ran the TU Campus El Gouna pilot project in Egypt for almost ten years with a master's programme, but this was discontinued during the pandemic due to a lack of paying students and a lack of funding from the founder, TU alumnus Samih Sawiris. [5] Many projects like this end when funding expires. I have rarely seen projects that went on to become successful self-sustaining ventures afterwards.

Without exception, the universities I visited have established contacts abroad, often in Europe, but also in the US, India, and China. Within Africa, universities seem to have significantly fewer connections. Is this because they hope to benefit more from contacts outside Africa? Studying or doing a doctorate in Europe continues to be beneficial for those who want to succeed. Many professors I met can attest to this. People also bring their knowledge back with them from where they did their doctorates, and it is not surprising that the teaching material and standards from abroad are often adopted more or less unchanged. Aside from South Africa and Egypt, there are practically no national standards; instead, existing standards such as the Eurocodes, the American AASHTO, or the British Standards are used. Adapting these standards to local conditions or developing independent standards is time-consuming and costly. However, directly adopting standards can also be expensive if their assumptions from other climate zones and labour markets, as well as high comfort and safety standards that make construction expensive and resource-intensive, are not questioned.

Africa is now entering a period of major construction activity. Houses, roads, and bridges urgently need to be built for the hundreds of millions of people who will be born in Africa in the coming decades. The civil engineers who will solve these tasks are currently being trained. They will have to deal with sustainability issues in particular. Of the countries I travelled to, only South Africa and Rwanda currently have a mandatory building energy code. [6] In Europe, on the other hand, such standards are now the norm. When new professorships in civil engineering are filled in Germany, their focus is increasingly on additive manufacturing, digitalisation, modular construction, and lightweight construction, and the titles of the professorships are expanded to include adjectives such as 'resource-efficient', 'sustainable', 'circular', 'climate-friendly', 'ecological', 'resilient', or 'environmentally friendly'. Even if 'green washing' is occasionally involved, serious attempts are now being made in teaching and research to address issues of sustainability and climate change. The mistakes we have made in construction over the last few decades are only now gradually leading to a (too) late change of course.

Will African civil engineering be able to avoid resource-wasting, environmentally damaging, and architecturally worthless construction? Certainly, many African universities lack the money for expensive laboratories, large computers, and 3D printers to tackle sustainable construction in research. I only met a few civil engineers who deal with issues of *Baukultur*, with structures that can also contribute formally to the quality of life. Discourses on sustainable construction, the development of buildings that make sense in terms of building physics and minimise the use of materials, the use of local and abundant natural materials for industrial use, or the translation of traditional construction methods into the present day are currently practically non-existent as research topics or part of teaching.

In his book *New Directions in African Architecture* [7], published in 1969, Udo Kultermann presents projects by the first architects trained in Africa after their countries gained independence and states: "it is not yet possible to speak of a truly African architecture". He describes their task as the difficult bridging of "between the day before yesterday and the day after tomorrow", rediscovering forgotten traditions, "lesser known local technologies" [8], and having an enriching effect worldwide. Since then, so much has happened in African architecture that Kultermann would probably omit the "not yet" today.

I see civil engineering in many African countries today in a similar situation to that of architecture more than 50 years ago. Can African civil engineers derive their own engineering culture from their context, the deserts and rainforests, the snow-capped mountains and coasts with their respective climates, soils, materials, and building traditions? I am not talking about a 'truly African civil engineering' style. Such a thing has never existed in civil engineering, either in Germany or elsewhere. But couldn't a leap to 'the day after tomorrow' also skip the mistakes we have made? Couldn't the new African civil engineering begin right where it concerns resource-saving structures made from local and regional building materials? For the production of which clean and inexhaustible solar energy is used? Can the new generation achieve this, and can they be supported in doing so?

Fig. 1 Visits to universities, institutions, and engineering firms

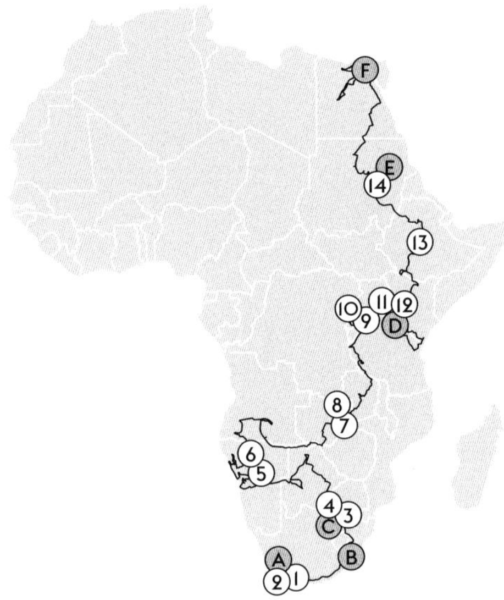

	Universities	Country	Date of visit/lecture	Civil engineering since approx.
1	Stellenbosch University	South Africa	16 March 22	1918
2	University of Cape Town	South Africa	29 March 22	1918
3	University of Pretoria	South Africa	22 March 22	1908
4	University of Botswana, Gaborone	Botswana	15 April 22	1982
5	NUST, Windhoek	Namibia	1 April 22	1996
6	UNAM, Ongwediva	Namibia	11 May 22	1992
7	UNZA, Lusaka	Zambia	24 May 22	1964
8	Copperbelt University, Kitwe	Zambia	27 May 22	1987
9	University of Rwanda, Kigali	Rwanda	8 June 22	1960
10	INES, Ruhengeri	Rwanda	9 June 22	2003
11	Makerere University, Kampala	Uganda	11 June 22	1970
12	Jomo Kenyatta University, Juja, Nairobi	Kenya	20 June 22	1994
13	Addis Ababa Institute of Technology, Addis Ababa	Ethiopia	22 July 22	1964
14	University of Khartoum, Khartoum	Sudan	3 August 22	1939
	Institutions and engineering firms			
A	SMEC, Cape Town	South Africa	17 March 22	
B	SMEC, Msikaba Bridge construction site	South Africa	11 April 22	
C	Nirox Foundation, Krugersdorp South Africa	South Africa	13 April 22	
D	SMEC, Nairobi	Kenya	17 June 22	
E	Ministry of Roads and Transportation, Khartoum	Sudan	4 August 22	
F	HBRC – Housing and Building National Research Center, Cairo	Egypt	28 August 22	

Source: Own representation

Solar Energy – Africa's Trump Card

Climate change, finite resources, and social responsibility: these issues are omnipresent and well known. In Africa, at least the question of clean and affordable renewable energy, which is available in abundance, could be posed and answered quite differently than in the northern hemisphere. Extensive deserts and strong solar radiation on the continent provide ideal conditions for harvesting large amounts of energy on suitable land and using it as electricity or in the form of hydrogen. I am convinced that the 'energy problem' will soon no longer exist.

There is growing global awareness that renewable energies must come from the sun, wind, and water in order to make transport, industry, and heating environmentally friendly and affordable. This is not possible with coal, oil, natural gas, or, at least in my opinion, nuclear fission, and nuclear fusion is still a long way off. That is why we can, indeed we must, advocate for the rapid and large-scale provision of desert electricity from solar thermal and photovoltaic (PV) power plants. The vast amounts of energy we all need can be produced particularly advantageously in the world's deserts. In areas that are not yet electrified, PV systems are also ideal for decentralised power supply. The technologies and transport options are available and proven, but political support is also needed to enable the necessary secure investments. The construction of solar power plants in desert areas could not only provide clean energy for local needs and opportunities to develop the domestic economy and exports, but also generate jobs. The sun can create livelihoods in Africa in two ways – ecologically and economically. Hopefully, this will be a win-win situation!

As long as the world is still dependent on fossil fuels, we are required to save energy, and we know that there is great potential for savings in all areas of energy consumption, from construction and heating to transport. The electrification of these areas promises efficiency gains, and by 2045, Germany's final energy consumption is expected to fall to 1,400 TWh/a (terawatt hours per year). The fact is, however, that according to the Federal Environment Agency, final energy consumption in Germany has hardly fallen so far and has remained constant at around 2,500 TWh/a for 30 years. [1] In 2022, around 46 per cent of the electricity Germans consume already came from renewable sources. [2] However, electricity only accounts for part of our energy consumption. Looking at all sectors, including heating and transport, the share of renewable energies in 2022 was just under 21 per cent of Germany's final energy consumption. [3] Furthermore, the report by the Federal Environment Agency for the windy and sunny year of 2022 shows that electricity from wind turbines and photovoltaics accounted for around

Fig. 1 Energy for 2.5 billion people in Africa in 2050

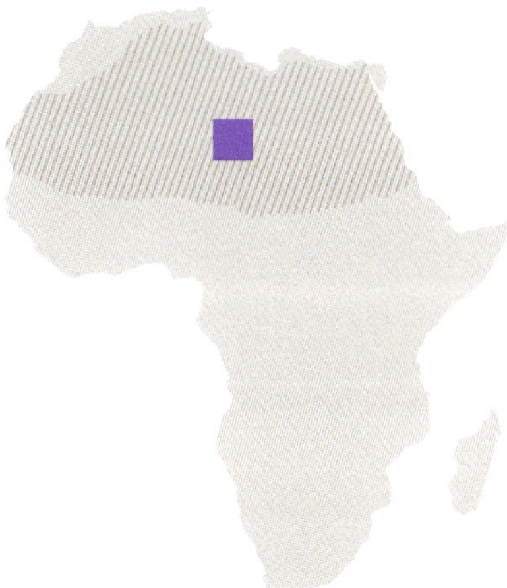

Total area of the Sahara
approx. 10,000,000 km²

2,2 %
approx. 220,000 km² ~ 470 x 470 km

Solar power plants covering this area could meet Africa's energy consumption in 2050

Source: Own representation

63 per cent of the renewable energies generated. Photovoltaics and wind turbines currently generate 0.63 × 0.21 = 13 % of Germany's final energy requirements. Assuming that our final energy consumption remains constant in the future and that the remaining 79 per cent of our energy consumption that is not renewable will be covered solely by wind turbines and photovoltaics in the future, we would need around six times as many wind and PV power plants as we have today.

In principle this is possible in Germany. Of course, the use of wind and PV energy in this country still needs to be expanded massively. Despite general acceptance of wind and solar power, there is also criticism. Namely, when there is a threat of the loss of valuable farmland or disfigured cityscapes and landscapes. Furthermore, to achieve true self-sufficiency so that green hydrogen and methane can cover seasonal shortages in all areas, far more than six times the solar and wind power plants would have to be built. We are still a long way from this 'power-to-gas' scenario in Germany, which is why, for the time being, in addition to importing fossil fuels, gas and coal-fired power stations must be maintained at great expense and nuclear power must be purchased from neighbouring countries.

In the desert regions of North Africa, the average solar radiation is around 2,300 kWh/m²a, which is about twice as high as in Germany, and huge areas of 'wasteland' are available there. Taking into account the irradiated energy per area E, PV efficiency η_{pv}, transmission efficiency $\eta_{ü}$, and ground coverage ratio GCR, solar power plants there can provide around $E \times \eta_{pv} \times \eta_{ü} \times GCR = 2{,}300\ kWh/m^2a \times 0.3 \times 0.90 \times 0.3 = 0.19\ TWh/km^2$ of final energy per year. To cover Germany's total energy requirements, around $2{,}500/0.19\ km^2 = 13{,}157\ km^2$ of desert area would be needed, which corresponds to 3.7 per cent of Germany's total area, a circle with a diameter of around 129 kilometres, or around 0.13 per cent of the area of the Sahara.

Africa's total final energy consumption in 2022 was only around 20 exajoules [3], or 5,555 TWh, which could be generated by solar power plants on just under 0.3 per cent of the Sahara's surface area. It must be taken into account that its energy consumption will rise sharply in the coming years. If Germany's final energy consumption were to fall to 1,400 TWh/a for its 85 million inhabitants, as planned for 2045, and we were to convert this to Africa with a population of 2.5 billion, Africa's annual consumption would be 1,400 × 2,500 / 85 TWh = 41,176 TWh. This would correspond to 2.2 per cent of the area of the Sahara to cover Africa's own needs Fig. 1.

Of course, electricity will not only come from the Sahara in the future. Africa and Europe need energy from different, complementary technologies and regions for security and independence. However, this simple calculation clearly shows Africa's potential for generating renewable energy. The continent offers 60 per cent of the world's best solar resources Fig. 2. [3]

Thirty-five years ago in August 1989, my father Jörg Schlaich called for the large-scale use of solar power and the construction of wind power plants in his essay *Wieviel Wüste braucht ein Auto ('How much desert does a car need?')* [4]. At that time, the electricity production costs of photovoltaics were many times higher than today at 2 Deutschmark/kWh (1€/kWh), and focussing solar thermal power plants were still in the research stage. Today, PV systems with a capacity of 100 MW and electricity production costs of less than 2 €ct/kWh are commonplace in sunny locations. Large PV power plants often use efficiency-enhancing 'trackers', rotating support structures that allow the PV panels to follow the sun. Large PV power plants are often combined with solar thermal concentrating power plants, which achieve similarly high-power ranges and are also considered 'proven technology'. Such CSP (concentrating solar power) plants first produce heat, which is easier to store than the electricity generated by PV systems. While the PV system generates inexpensive electricity during the day, the CSP plant collects heat for the production of valuable electricity that is used at night. Trough-type CSP power plants and tower-type CSP power plants are currently being built. In the former, kilometre-long parabolic troughs with mirrored surfaces concentrate sunlight onto tubes located in their focal line, through which a heat transfer medium (oil or liquid salt) flows. The medium, heated to around 400° Celsius in the case of oil, can then be used via heat exchangers to directly drive steam turbines for electricity generation or to heat salt storage tanks, whose heat can be accessed as needed. Significantly higher temperatures and thus better efficiency can be achieved with tower power plants. In these CSP plants in the 100 MW range, thousands of individual mirrors called 'heliostats' reflect the sun's rays onto a focal point, the receiver, which is mounted on towers over 150 metres high. In the receiver, which is a system of pipes the height of a building carrying liquid salt, temperatures of about 600° Celsius are reached. In addition to the high temperatures, the advantage of these plants is the short distance that the heat transfer medium has to travel, namely only from the storage tank at the foot of the tower to the receiver. On the other hand, the many precision mirrors that track the sun and have to hit the receiver from a distance of hundreds of metres are more complex than in trough power plants.

One example of such a combined power plant is Noor Ouarzazate in Morocco, with a total capacity of more than 500 MW [5]. Two trough power plants and a tower power plant with a 250-metre-high tower, all with large thermal storage facilities that can keep the power plant running for up to 7.5 hours without sunlight, are supplemented by a 72 MW PV field. The planned Noor Midelt plant [6] with 800 MW combines CSP with a higher PV share. Morocco is Africa's pioneer in the field of solar energy and impressively demonstrates the potential that these technologies already offer today. Several large solar power plants are now also connected to the grid in Egypt, led by the Benban solar park near Aswan with a capacity of 1,659 MW. The 37-square-kilometre PV power plant, Africa's largest, went online in 2019 and covers 3 per cent of Egypt's energy consumption. [7] South Africa has five CSP power plants, each with a capacity of

100 MW [7], and is currently building the Kenhardt Solar Power Complex Station, a PV power plant with a capacity of 540 MW [8].

Namibia has particularly ambitious plans, with a power plant to be built near Lüderitz by the Namibian company Hyphen Hydrogen Energy with German participation. This plant for the production of electricity, hydrogen, and ammonia using PV and wind power is expected to have a final output of no less than 5 GW. The cost of US$10 billion is almost equivalent to the country's GDP, which shows that it only has a chance of success with massive borrowing or the involvement of external partners. [3] If the costs are kept within budget, this plant would still be relatively inexpensive: in the United Kingdom, the Hinkley Point C nuclear power plant with a capacity of 3.26 GW is currently being built for around €50 billion.

Kenya, though, is currently the model African country in the field of renewable energies. The country already covers almost 90 per cent of its electricity consumption with renewable energies, using solar, wind, and hydro power, but above all geothermal energy from the East African Rift Valley. The first power plant was commissioned in Olkaria in south-western Kenya in 1981 to drive turbines for electricity generation using hot steam. Geothermal power plants with a capacity of almost 900 MW have now been installed. [9] They produce around 40 per cent of Kenya's electricity [10], which also illustrates how little electricity the 50 million inhabitants currently consume. Kenyan PV power plants currently contribute only 1 per cent of electricity production. This shows that there is still great potential here to supply places that are not connected to the national grid with stand-alone solutions.

In the other African countries I have visited, solar and wind power as well as geothermal energy play virtually no role Fig. 3. Angola continues to rely on its large oil reserves, and in Zambia, hydropower has been sufficient so far, although a drought in 2019 reduced the output of the country's important Kariba Dam to 10 per cent. [3]

Throughout my journey, I have seen valuable trees being felled to produce charcoal and people using wood for cooking and heating in huts without flues. Due to frequent power cuts, generators that produce expensive and dirty electricity using diesel are everywhere. Solar power can provide an affordable, healthier, and more environmentally friendly solution in all these places.

The sun can enable Africa to supply itself with clean energy and become an energy exporter. The high level of solar radiation makes it easier than in Europe to provide solar power for industry and to produce hydrogen. This allows steelworks and even cement factories to be operated sustainably on site, as demonstrated by a pilot project by cement manufacturer Cemex. [11] Desert sand can become valuable because it can be converted into glass [12] and foam glass as an aggregate for concrete using solar power. High-voltage direct current lines or ships for hydrogen, ammonia, or methane from solar power are proven technologies for energy transport to Europe.

The sun and deserts are the foundation for Africa's future. The construction and operation of solar power plants creates jobs and reduces the need for migration. Do we see the development policy dimension of solar power plants in Africa? How many refugee camps and reception centres in North Africa will become redundant if, instead, investments are made in solar power plants and infrastructure that secures the foundations for a dignified life?

Fig. 2 Distribution of global solar radiation

Long-term average of direct normal radiation
in kWh/m²

Daily totals:	4	5	6	7	8	9	10
Annual totals:	1,461	1,826	2,191	2,556	2,922	3,287	3,652

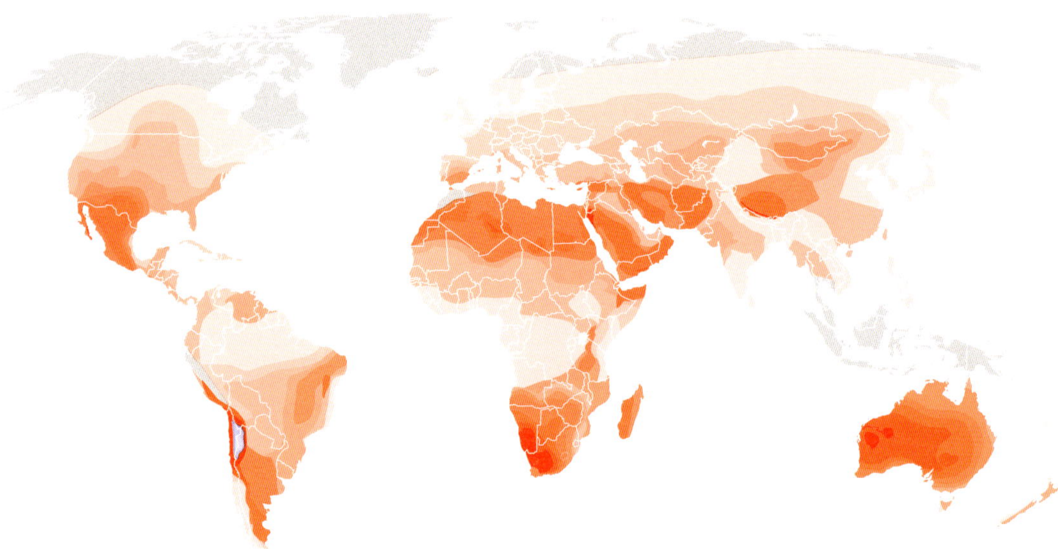

Source: Representation based on World Bank Group, Solargis, Solar Resource Map, Direct Normal Irradiation, 2023

Fig. 3 Renewable energies in Africa

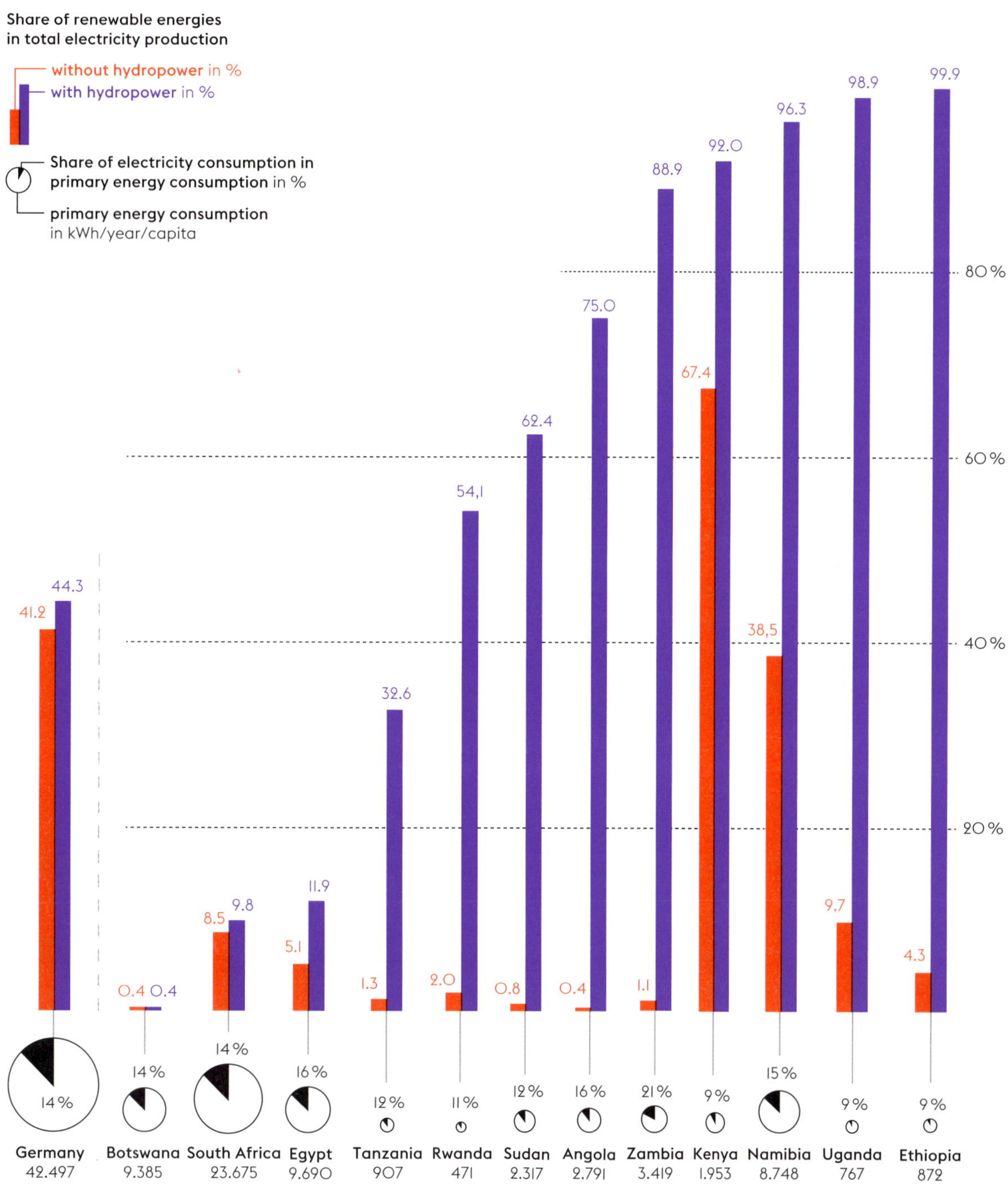

Source: Our World in Data | Wikipedia, see data sheet from page 220 onwards

Concrete – A Natural Building Material in Abundance

1

2

3

Figs. 1–3 Construction phases of a chika house

Concrete, which is currently the subject of criticism, could be the building material of the future for Africa. Concrete consists of cement, water, and aggregates. Foam glass and expanded glass made from desert sand using solar energy could replace the increasingly scarce aggregates. Cement can also be produced in a more environmentally friendly way than before using solar energy. There is no shortage of desert sand and sunshine in Africa.

Traditionally, houses in Africa are built locally using regional materials. All over the countryside, small, beautiful, and sustainable houses are built with wood, stone, clay, and grass, adapted to local conditions. This tradition will certainly continue, but in Africa's megacities, 'urban villas' built in this way will reach their limits.

In Ethiopia, new settlements with corrugated iron-roofed chika houses can be found at every turn. Their walls are made of 'wattled' eucalyptus wood, sealed with clay and straw and coated with paint to protect them from the rain Figs. 1–3. Eucalyptus has been cultivated on a large scale in Ethiopia since the late nineteenth century as a building material and as firewood. This fast-growing and relatively undemanding tree is the subject of controversy because it requires a lot of water and, when cultivated industrially, can lead to a loss of biodiversity and soil degradation due to its allelopathic properties. [1] Furthermore, timber harvesting does not appear to be sustainable everywhere. During our trip, we repeatedly saw large areas of cleared mountain slopes that were exposed to erosion. Over the course of the twentieth century, the area covered by dense forests in Ethiopia has declined from around 40 per cent to 3 per cent of the country's land area [2], and the situation is no different in many other African countries. Massive reforestation programmes with eucalyptus, cypress, and pine trees have increased Ethiopia's forest cover to around 15 per cent again [3], but this is hardly sufficient for the construction tasks ahead Fig. 4.

Around ten tonnes of eucalyptus poles are used to build the walls of a chika house with a floor area of 58 square metres. [4] Assuming that around 500,000 such houses need to be built annually in urban areas in Ethiopia alone [5], this would require 5 million tonnes of wood per year. That is more than the country's total roundwood production of just under 3 million tonnes (2018). [6] Should Ethiopia really plant that much eucalyptus? And wouldn't it be better to leave the trees standing so that they can produce oxygen, and build with concrete instead? Werner Sobek comes to this conclusion, at least for reinforced concrete columns, in his book *non nobis*. [7] Concrete has been used for construction throughout Africa since its global industrial emergence over 100 years ago. There are cement plants in almost every country in Africa. Everywhere in the countryside, you can see places where hollow blocks with low material consumption and relatively good thermal insulation or cement-reinforced clay masonry are produced. Concrete is freely mouldable, waterproof, durable, and inexpensive. Assuming that the exterior wall area of a chika house measures 100 square metres and adding 30 per cent for the interior walls, if everything is constructed with 20-centimetre-thick concrete walls, this would amount to 26 cubic metres of concrete. For 500,000 houses, this corresponds to 13 million cubic metres of concrete. Assuming 300 kilograms of cement per cubic metre of concrete, this would consume 3.9 million tonnes of cement. That is less than 30 per cent of the country's cement plant capacity in 2022 (see table at the end of the book), which means it is easily achievable in terms of cement. The rest of the concrete volume is made up of mixing water and aggregates.

However, concrete has come under heavy criticism recently. One reason for this is that the aggregates required in addition to cement – gravel, crushed stone, and sand – are becoming so scarce that in some countries gravel mafias are stealing river gravel from nature reserves. Desert sand, which is available in abundance throughout Africa, is generally suitable as an aggregate for concrete. However, if it is too fine [8], it must be mixed with coarser aggregates to achieve the appropriate grading curve, i.e. the grain mixture required for concrete. Desert sand may need to be desalinated before it can be used as an aggregate. Desert sand can also be further processed: even the pharaohs used to make glass by heating desert sand. This is because desert sand often consists of quartz, silicon dioxide (SiO_2), the main component of glass. Ground glass can in turn be heated to produce foam glass and expanded glass. [9] The energy required to produce glass and expanded glass can come from the sun. The above-mentioned quantities of aggregates certainly do not pose a threat to Africa's vast deserts. Sustainably sourced aggregates made from glass and expanded glass in all possible grain sizes and shapes for concrete production, therefore, do not appear to be a problem.

On the other hand, CO_2 emissions from cement production are considered a problem. In fact, cement production is responsible for an enormous 6 to 7 per cent of global CO_2 emissions. [10] Around 40 per cent of this is attributable to fossil fuels, which are needed to heat the raw materials clay and lime for conversion into clinker (cement is ground clinker). However, the fossil fuels, often old tyres, can be replaced by solar power. The remaining 60 per cent is attributable to the chemical process of converting clay and lime into clinker, known as 'deacidification'. As the name suggests, this process produces carbon dioxide or CO_2. This calculation often overlooks the fact that concrete – once poured and hardened – begins to carbonate. At least light and porous concretes can reabsorb up to 40 per cent of the CO_2 produced through deacidification during this process. [10]

So, if solar power is used in cement production and carbonation is accounted for, the carbon footprint of cement could be reduced by up to 64 per cent.

At the Technische Universität Berlin, we have been working on infra-lightweight concrete (ILC) as a load-bearing, durable thermal insulation material for almost 20 years. [11] This porous concrete weighs only 600 to 800 kg/m³, less than a third of normal concrete. We use expanded clay and expanded glass, which can be produced from waste glass, as lightweight aggregates. ILC can simultaneously bear the loads of multi-storey buildings and serve as thermal insulation. It is not suitable for bridge construction. It carbonates completely and, because it does not need to be bonded with additional thermal insulation, it is easy to recycle. We are, therefore, currently conducting research at the TU Berlin, together with Mahmoud El Shahawy from Egypt, to determine whether the desert sand there is suitable for the production of expanded glass. As part of another research project, research assistant Berhanu Mengistie from Ethiopia is investigating how the abundant pumice from the East African Rift Valley can be used as a lightweight aggregate for lightweight and infra-lightweight concretes to replace expanded glass and expanded clay with this natural material of volcanic origin Figs. 5–7. In Ethiopia's Amhara region alone, several hundred million cubic metres of pumice can be mined. [12] Approximately 500 kilograms of pumice aggregate (roughly equivalent to 0.5 cubic metres) are required for 1 cubic metre of ILC. For our chika house with 26 cubic metres of concrete, 13 tonnes of pumice are, therefore, required as the sole aggregate. The 6.5 million tonnes of pumice (= 6.5 million cubic metres) required for the 500,000 houses that Ethiopia needs each year could be extracted regionally. Adding finely ground pumice can also save up to 30 per cent of cement. [13] Replacing the chika house made of eucalyptus with infra-lightweight concrete with pumice aggregates is, therefore, worth considering. In my opinion, such approaches in concrete construction can be transferred to many countries in Africa. Nevertheless, the quantities of concrete required for building construction throughout Africa in the coming decades will be enormous. In any case, the supporting structures in building construction must be designed in such a way that concrete consumption is minimised. Once again, on the subject of carbonation: it should not be concealed that it leads to faster corrosion of the steel reinforcement commonly used today. For this reason, galvanised steel reinforcement made of glass, basalt, or carbon fibres is certainly advisable for highly carbonating concretes. At least glass and basalt are available in abundance. Once again, renewable solar energy can be used to produce fibres from these materials.

Designing and constructing with concrete, used consciously and minimally, can, as described here, make an important contribution to environmentally friendly construction in Africa.

5

6

7

Figs. 5–7 Three types of pumice from Ethiopia

Fig. 4 Forest area and roundwood trade in Africa

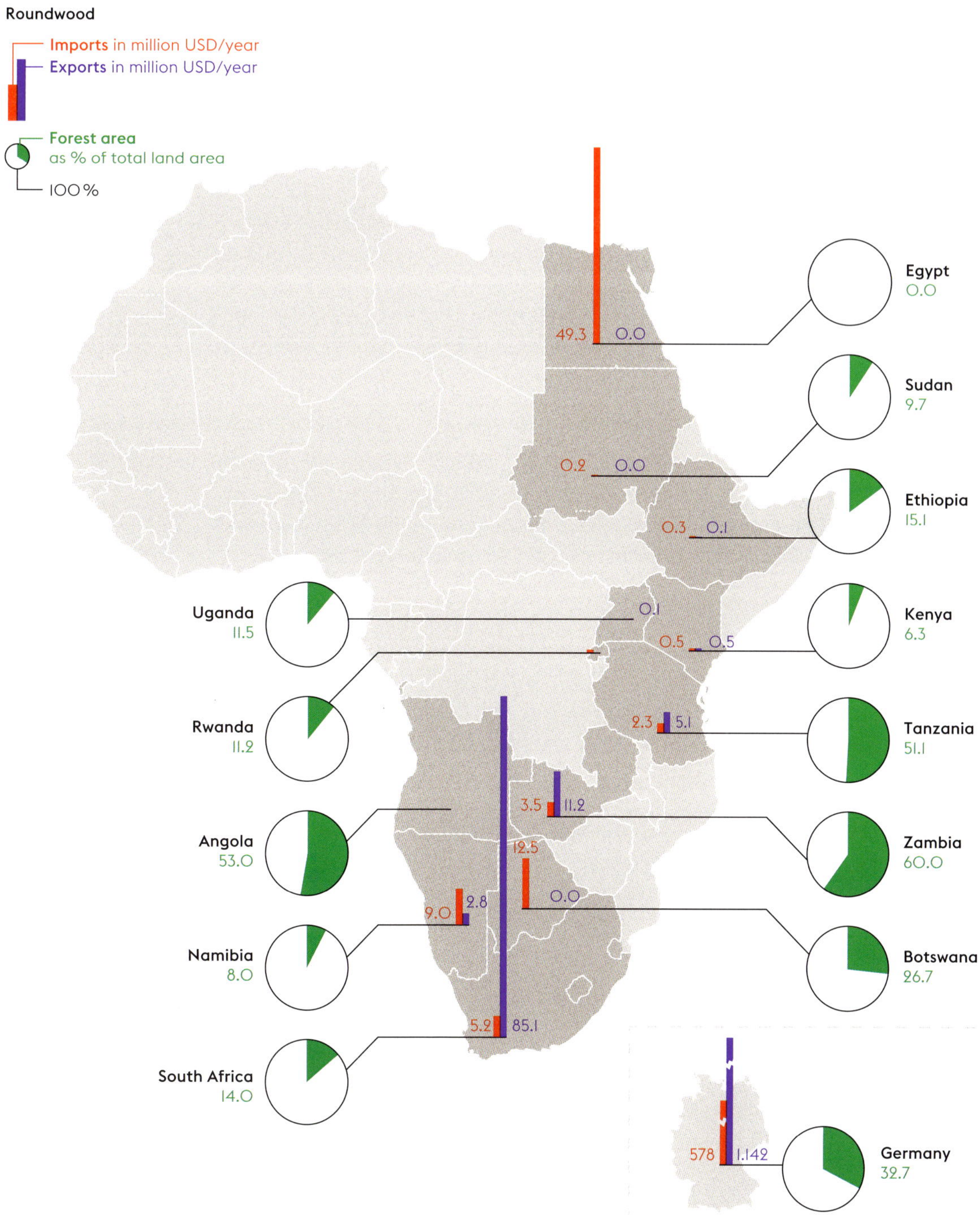

Source: World Bank | UN Comtrade Database, see data sheet from page 220 onwards

Bridge Building – A Wonderful Challenge

In sub-Saharan Africa, the masonry arch bridges built near Gondar in Ethiopia in the seventeenth century, some of which are still in use today, are among the oldest surviving bridges (see also diary). Resembling European stone bridges of that period, they are also called 'Portuguese bridges' because Jesuits from Portugal were in the country at the time and may have contributed their knowledge. The white settlers from the Netherlands and the Huguenots, who arrived in the Cape Town region from 1650 onwards, soon moved into the mountainous interior. By the middle of the eighteenth century, local administrations were already building mountain passes and simple bridges made of wooden beams and stone slabs. In his informative and inspiring book *The Romance of Cape Mountain Passes* [1], Graham Ross describes South Africa's first stone arch bridge. In 1825, 150 soldiers from the Royal African Corps built Jan Joubert's Gat Bridge on the Franschhoek Pass with a 5-metre-wide arch. It is still open to traffic today. With the advent of the railway and the automobile, European colonialists also brought bridges to Africa in order to open up the interior of the country from the coasts. The steel for bridge construction initially came from Europe and in southern Africa it came from South Africa from the twentieth century onwards. This led to the construction of railway bridges throughout Africa made of prefabricated, riveted trusses or solid web girders, from the historic railway bridge near Windhoek to the Old White Nile Bridge in Khartoum, Sudan (see also diary). To this day, Europe exports bridge design and construction to Africa, helping to open up the continent. While forced labourers were often exploited for construction during colonial times, our export products today hopefully also support local efforts to train skilled workers, create jobs, and provide access to safe transport routes. Africa's longest suspension bridge at present is the Maputo-Katembe Bridge in Mozambique, completed in 2018 with a span of 680 metres. With a main span of 580 metres, the Msikaba Bridge currently under construction in South Africa will be the longest cable-stayed bridge on the continent. The Bloukrans Bridge mentioned below holds the span record for arch bridges, while the 6 October Bridge in Cairo holds the length record at around 20 kilometres.

South Africa, with its long history of bridge building, has produced a whole series of remarkable bridge structures. There does not seem to be a book dedicated solely to bridge building in South Africa, but Wikipedia lists the most important bridges with references. [2] The best known are probably the beautiful large bridges along the N2 national road in the Western Cape province, above all the 450-metre-long Bloukrans Bridge by Charles Liebenberg, completed in 1983, with its 270-metre-long reinforced concrete arch. A number of footbridges are also worth a visit, such as the simple suspension bridges over the Storms River in Tsitsikamma National Park, the curved treetop walkway in Kirstenbosch Botanical Gardens in Cape Town [3], and the motorway overpasses designed by the South African National Roads Agency (SANRAL) [4] to fit in with the local context. South Africa is home to a number of large and long-established engineering firms with experience in bridge and infrastructure planning, although not all of them are still South African-owned. The most important ones include:

→ Stewart Scott Ltd, founded in 1922, now Royal HaskoningDHV, Amersfoort, Netherlands;
→ Ninham Shand, 1932, Connell Wagner, 1934, and Africon, founded in 1951, now Zutari, Pretoria, South Africa;
→ Hill Kaplan, founded in 1956, now GIBB, Johannesburg, South Africa;
→ BKS, 1965, now AECOM, Dallas, USA;
→ VKE Consulting Engineers, 1982, now Surbana Jurong Group, Singapore.

At the northern end of the continent, Egypt represents another centre of African bridge construction. Interestingly, despite their highly developed building culture, the pharaohs had little interest in permanently bridging the Nile. Civil engineer Mourad Bakhoum from Cairo, who has studied the country's bridge-building history extensively, suspects that the pharaohs considered crossing the river with powerful ships and the occasional construction of temporary pontoon bridges made of barges to be sufficient. He cites the Qasr El Nil Bridge in Cairo, built in 1871 and replaced in 1930, as the first Egyptian bridge over the Nile. [5] He refers to an aqueduct in Cairo with pointed arches dating from the time of Sultan an-Nasir Muhammad ibn Qalawun in the fourteenth century and to mud brick arches in buildings that must have been built 3,500 years ago. Today, Egypt has more than 50 bridges across the Nile and three cable-stayed bridges – two across the Nile in Aswan and Cairo and one across the Suez Canal. [6] A number of Egyptian engineering firms with experience in large-scale national and international bridge construction, such as ACE (Mourad Bakhoum), Sabri Samaan Consulting, SICE (Fathy Saad), and NECB (Farouk El-Kadi), are capable of undertaking major planning tasks independently.

Outside of Egypt and South Africa, there are currently few African engineering firms and construction companies with many years of bridge-building experience. Bridge planning and construction must continue to be imported there. Conventional road and rail bridges with smaller spans often come from China today, as I observed on my trip. The Japanese, South Koreans, and Europeans are involved in more sophisticated long-span bridges. Who gets the contract depends not least on the country of origin of the financier or lender. Nicolas Janberg's website *Structurae* offers a nice overview of the diversity of bridge construction. [7] No fewer than 600 bridges from across Africa are listed there with additional information such as photos, bridge types, and locations. Nearly

200 of these are 'trail bridges', such as those built by the organisation Bridges to Prosperity [8, 9], which builds such bridges in rural East Africa with the involvement of the local population.

Small footbridges and small houses are built all over the world by local communities themselves. The diversity of this construction is the result of centuries of experience. It varies from region to region, just as the climate and local building materials vary. Building has always been expensive, and transport has always been laborious and costly. That is why local self-building also means building with local materials. The situation is quite different for modern railway and road bridges. They must be able to withstand heavy loads, high dynamic forces, strong winds, and earthquakes on a permanent basis. This applies all over the world. Bridge standards prescribe similar rules for design and construction everywhere. This is a good thing, meaning that we can rely on comparable load-bearing capacity, safety, and durability worldwide. Local self-building is rarely possible for large bridges.

However, this does not mean that a local or national bridge-building culture cannot develop. Local conditions for bridge construction, such as rain and flood risks, salt content in the air, snow, wind and earthquake loads, vary from place to place. Building materials can differ in terms of cost, but also in terms of strength, durability, and colour. The ratio of road to pedestrian traffic and their speeds can result in different bridge widths and traffic separations. National standards can be adapted flexibly and in a way that promotes innovation to take all of this into account. Last but not least, there are cultural differences and diverging ideals of beauty. The context can be as varied as the solutions for designing a bridge.

Above all, however, there is no reason to repeat in Africa the mistakes we have made in bridge construction in the Global North. Especially in the case of conventional bridges with smaller spans, I often see heavy single-span girders made of precast concrete elements in Africa, which, due to their many joints and bearings, can be neither durable nor slim and material-saving. The design and construction of these inexpensive bridges often seem to have been imported without taking the context into account. Unfortunately, such bridges, strung together in long elevated roads with identical spans, also spoil the centres of African cities such as Nairobi and Addis Ababa, or they are an eyesore in the open countryside. Of course, there is no panacea, but becoming aware of the difference between cost and value and not always hiring the cheapest design and construction company would be a first step towards more careful planning and higher-quality construction and towards a better bridge-building culture. Fewer single-span bridges and more joint- and bearing-free bridges would lead to less maintenance and greater durability. Single-span bridges can be combined to form lightweight monolithic structures and integral bridges. Material-appropriate and knowledgeable design and construction with sensibly combined materials can lead to resource-saving lightweight construction. Taking environmental concerns into account during the planning stage and conducting sustainability studies will prevent damage in the future and many of the mistakes we have made. There are still a number of obstacles to bridge construction in Africa. To begin with, national standards as well as earthquake and wind load maps are lacking everywhere, which means that international standards have to be used. For small and large, modest and prestigious bridges alike, there is a need for well-trained African bridge engineers who are able to design and construct in accordance with the local context. One suggestion would be for experienced international engineering firms to invite African colleagues to undergo further training for a certain period of time. Not so that our way of building is imitated, but so that we can learn from each other and also establish contacts for future cooperation.

Europe's infrastructure is largely complete. Therefore, the focus there is increasingly on repair, reinforcement, or replacement. This is not the case in Africa, where many new bridges still need to be built. At the Technische Universität Berlin, we used Google Maps to count the permanent crossings (dams, road and railway bridges) over the four longest rivers in Africa (the Nile, the Congo, the Niger, and the Zambezi), which together are 18,000 kilometres in length. For the Nile – from Lake Victoria (White Nile) and Lake Tana (Blue Nile) to the Mediterranean Sea – we counted 115 fixed crossings; for the Congo it was 5; it was 23 for the Niger and 14 for the Zambezi Fig. 1. In total, there are only around 157 places where you can cross these four rivers without getting your feet wet. The four longest rivers in the EU (the Danube, the Rhine, the Elbe, and the Loire) have a combined length of around 6,000 kilometres, with a bridge on average every ten kilometres. Around 190 bridges cross the approximately 1,200-kilometre-long Rhine. [10] We counted 251 bridges along the 2,857-kilometre-long Danube. Applied to Africa, whose population density will soon match that of Europe, this would mean that around 1,800 large bridges would be needed on the four longest rivers alone, around 12 times as many as currently exist. Perhaps we will not need as many bridges in the future as we do today; perhaps a ferry will sometimes suffice. And perhaps decentralised structures are suitable for reducing mobility in general. In any case, bridge construction offers civil engineers in Africa the opportunity to go their own way.

In recent years, we have offered a series of lectures on *Brückenbaukultur* (*bridge building culture*) at the Institute of Civil Engineering at TU Berlin, where civil engineers from Germany present their bridges. In the winter semester of 2023/2024, the topic was 'Brückenbaukultur in Afrika'. And indeed, seven German engineers who had worked on bridges in Africa were quickly found. They gave our students fascinating insights into the practice of bridge construction in Africa and showed us that bridge construction there can be both a pleasure and a burden. The entire spectrum of cooperation with Africa was represented: firms such as GAUFF and INROS LACKNER each have several branches in Africa and have been represented there for decades. Others, such as GRASSL and SSF, are in the process of establishing themselves. Still others have had to withdraw from conflict zones, as

Fig. 1 The bridges over the four longest rivers in Africa

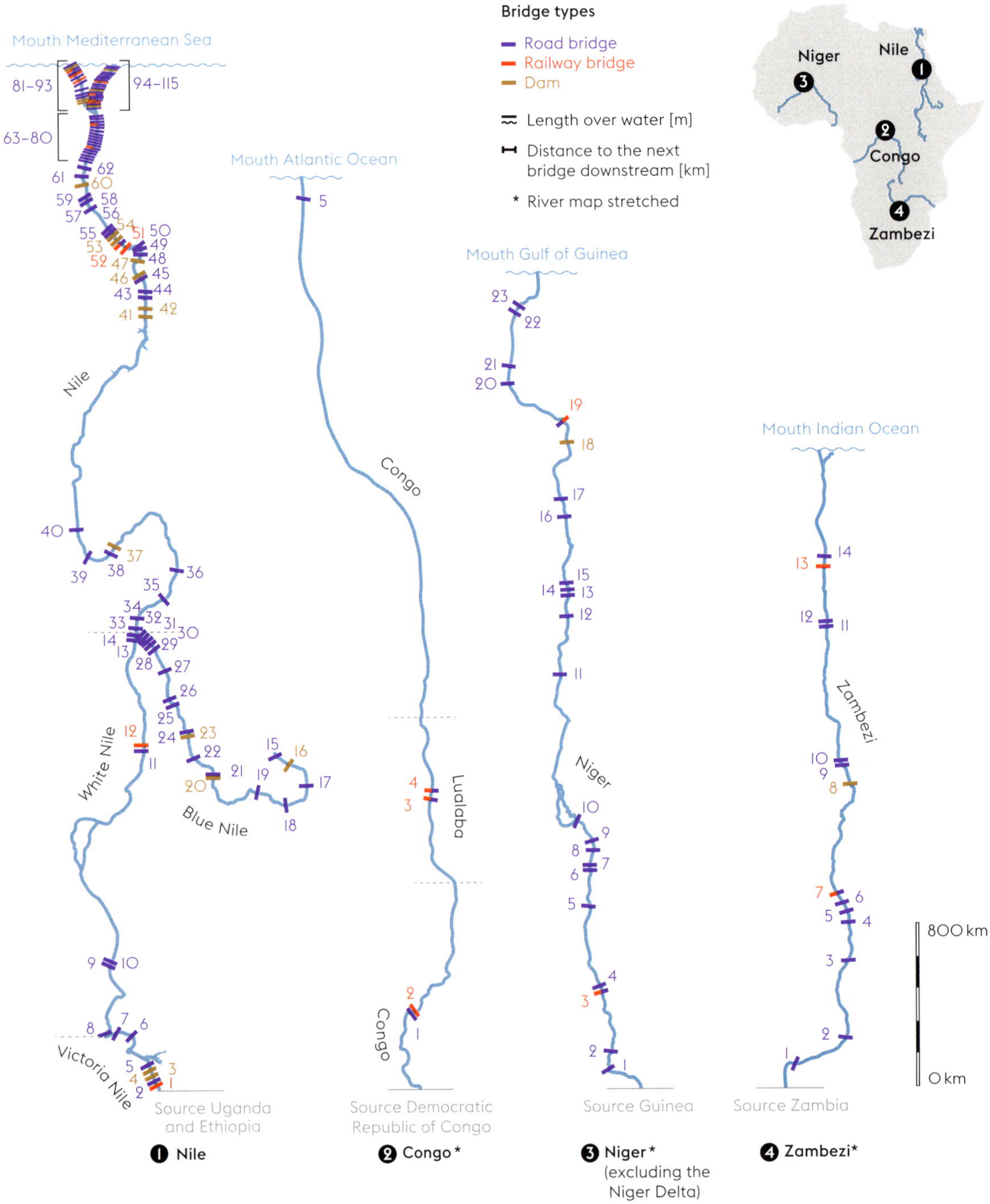

Source: Own representation based on Google Maps, April 2024

❶ Nile 6,650 km

Victoria Nile

1 Jinja Railway Bridge ≂ 105 ↔ 0.7
2 Source of the Nile Bridge ≂ 364 ↔ 0.5
3 Owen Falls Dam ↔ 42
4 Isimba Hydroléléctric Power Station ↔ 0.5
5 Isimba Bridge ≂ 262 ↔ 275
6 Karuma Bridge ≂ 82 ↔ 95
7 Paraa Bridge ≂ 345 ↔ 55

White Nile

8 Pakwach Bridge ≂ 160 ↔ 350
9 Freedom Bridge ≂ 400 ↔ 2
10 Juba Bridge ≂ 236 ↔ 1,216
11 Kusti Bridge ≂ 560 ↔ 0.8
12 Kusti Railway Bridge ≂ 465 ↔ 309
13 Al Fatehab Bridge ≂ 540 ↔ 1
14 Old White Nile Bridge ≂ 175 ↔ 4

Blue Nile

15 A3 Blue Nile Bridge ≂ 159 ↔ 35
16 Tisabay hydropower projekt II ↔ 171
17 B21 Abay Bridge ≂ 100 ↔ 101
18 A3 Blue Nile River Bridge ≂ 150 ↔ 178
19 Road 32 Blue Nile Bridge ≂ 60 ↔ 534
20 Roseires Dam ↔ 0.3
21 Roseires Damazin Bridge ≂ 300 ↔ 197
22 Sinjah Bridge ≂ 255 ↔ 96
23 Sennar Dam ↔ 6.0
24 Sinnar Bridge ≂ 280 ↔ 171
25 Hantoub Bridge ≂ 241 ↔ 53
26 Al Hasaheisa Rufaa Bridge ≂ 250 ↔ 130
27 Soba Bridge ≂ 302 ↔ 16
28 Manshia Bridge ≂ 279 ↔ 5
29 Kobar Bridge ≂ 255 ↔ 1
30 Old Blue Nile Bridge ≂ 352 ↔ 1
31 Al Mak Nimir Bridge ≂ 382 ↔ 2
32 Tuti Bridge ≂ 170 ↔ 4

Nile

33 Schambat Bridge ≂ 470 ↔ 8
34 Al Halfaia Bridge ≂ 650 ↔ 171
35 Shendi Almatma Bridge ≂ 366 ↔ 140
36 Qureir Arbara Road Bridge ≂ 500 ↔ 445
37 Merowe Dam ↔ 35
38 Marwi Friendship Bridge ≂ 379 ↔ 122
39 Aldabba Bridge ≂ 280 ↔ 152
40 Dongola Elsilaim Bridge ≂ 640 ↔ 759
41 Aswan High Dam ↔ 8
42 Old Aswan Dam ↔ 19
43 New Aswan City Bridge ≂ 516 ↔ 49
44 Kalabsha Bridge ≂ 672 ↔ 46
45 Edfu Bridge ≂ 380 ↔ 51
46 Old Esna Barrage ↔ 1
47 New Esna Barrage ↔ 46
48 Luxor Bridge ≂ 413 ↔ 44
49 Pont Qus Naqada ≂ 427 ↔ 28
50 Dandra Bridge ≂ 673 ↔ 2
51 Qina Railway Bridge ≂ 521 ↔ 53
52 Naga-Hammadi Bridges ≂ 351 ↔ 13
53 Old Naga-Hammadi Dam ↔ 3
54 New Naga-Hammadi Dam ↔ 44
55 Gerga Dar El Salam Bridge ≂ 475 ↔ 34
56 Sohag Bridge ≂ 818 ↔ 2
57 Akhmim Sohag Bridge ≂ 603 ↔ 49
58 Tama Bridge ≂ 574 ↔ 45
59 Assiut Bridge ≂ 374 ↔ 4
60 Assiut Barrage ↔ 63
61 Dayrout Bridge ≂ 428 ↔ 23
62 Mallawi Bridge ≂ 398 ↔ 51
63 Minya Bridge ≂ 340 ↔ 26
64 Samalut Bridge ≂ 326 ↔ 21
65 Beni Mazar Bridge ≂ 271 ↔ 74
66 Beni Suef Bridge ≂ 529 ↔ 5
67 Al Wasta Bridge ≂ 460 ↔ 71
68 Regional Ring Road Bridge ≂ 341 ↔ 12
69 Al-Marazeek Bridge ≂ 550 ↔ 7
70 Badrashin Corniche – Middle Ring Road ≂ 398 ↔ 16
71 Al Munib Bridge ≂ 656 ↔ 2
72 Abbas Bridge ≂ 394 ↔ 1
73 Cairo University Bridge ≂ 456 ↔ 1
74 Qasr El Nil Bridge ≂ 462 ↔ 0,5
75 6th of October Bridge ≂ 271 ↔ 1
76 15 May Bridge ≂ 260 ↔ 1
77 Imbaba Bridge ≂ 405 ↔ 1
78 Rod El Farag Bridge ≂ 467 ↔ 1
79 Tahya-Misr and Rod-El-Farag West Bridge ≂ 500 ↔ 5
80 El-Warraq Ring Road Bridge ≂ 780 ↔ 10

Rosetta Branch

81 Old Alkanater Kheirya Bridge ↔ 0.8
82 New Alkanater Kheirya Bridge ↔ 41
83 Bridge Bani Salama ≂ 196 ↔ 26
84 Tamlay Bridge ≂ 185 ↔ 51
85 New Kafr El Zayat Bridge ≂ 312 ↔ 0.2
86 Old Kafr El Zayat Bridge ≂ 396 0,.4
87 Kafr El Zayat Bridge ≂ 256 ↔ 49
88 Upper Desouk Bridge ≂ 370 ↔ 2
89 Old Desouk Bridge ≂ 326 ↔ 12
90 Fuwa Bridge ≂ 467 ↔ 14
91 Edfina Bridge ≂ 315 ↔ 2
92 Edfina Barrage Bridge ↔ 15
93 Rasheed Bridge ≂ 342

Damietta Branch

94 Mohamed Ali Bridge ≂ 385 ↔ 0.3
95 New Alkanater Kheirya ↔ 3
96 Alkanater Kheirya Bridge ≂ 210 ↔ 51
97 Banha Railway Bridge ≂ 240 ↔ 362
98 Kafr Al Gazar Bridge ≂ 264 ↔ 2
99 Banha Elevated Bridge ≂ 144 ↔ 2
100 Banha Regional Ring Road Bridge ≂ 212 ↔ 34
101 Mit Ghamr Zefta Bridge ≂ 204 ↔ 2
102 Mit Ghamr Tanta Bridge ≂ 256 ↔ 1
103 Zefta Barrage Bridge ↔ 34
104 Samannoud Bridge ≂ 164 ↔ 16
105 Mansoura University Bridge ≂ 185 ↔ 3
106 Talkha Bridge ≂ 210 ↔ 0.8
107 Mansoura Railway Bridge ≂ 192 ↔ 6
108 Schrangasch Bridge ≂ 195 ↔ 19
109 New Sherbin Bridge ≂ 168 ↔ 0.8
110 Old Sherbin Bridge ≂ 250 ↔ 38
111 Faraskur Bridge ≂ 268 ↔ 11
112 International Coastal Road Bridge ≂ 248 ↔ 7
113 Damietta Dam Bridge ↔ 3
114 Damietta Rotating Bridge ≂ 133 ↔ 0.8
115 Damietta Bridge ≂ 158

❷ Congo 4,835 km

1 N1 Bridge Bukama ≂ 70 ↔ 4
2 Pont Bukama ≂ 228 ↔ 543
5 Pont Matadi ≂ 520

Lualaba

3 R361 Bridge Kabalo ≂ 458 ↔ 91
4 Kongolo Bridge ≂ 440 ↔ 2,694

❸ Niger 4,184 km (excluding Niger Delta)

1 N1 Bridge Faranah ≂ 30 ↔ 57
2 N29 Bridge Balandougou ≂ 65 ↔ 205
3 Karoussa Bridge ≂ 160 ↔ 30
4 N1 Bridge Koumana ≂ 175 ↔ 81
5 N6 Bridge Niandan Diéliba Koro ≂ 446 ↔ 250
6 King Fahd Bridge ≂ 752 ↔ 1
7 Martyrs Bridge ≂ 753 ↔ 7
8 3ème pont de Bamako ≂ 1,115 ↔ 55
9 Pont de Kayo de Koulikoro ≂ 970 ↔ 200
10 Barrage de Markala ≂ 810 ↔ 1,088
11 Pont Wabaria Gao ≂ 280 ↔ 363
12 Pont Djibo Bakary ≂ 360 ↔ 56
13 Pont Général Seyni Kountché ≂ 479 ↔ 5
14 Kennedy Bridge ≂ 420 ↔ 2
15 Pont de Chinois ≂ 261 ↔ 294
16 Pont de Malanville ≂ 295 ↔ 117
17 Tuga Bridge ≂ 700 ↔ 210
18 Kainji Dam ↔ 98
19 Jebba Road and Railway Bridges ≂ 1,340 ↔ 288
20 Confluence Bridge Abuja Lokoja ≂ 1,005 ↔ 71
21 Ajaokuta Itobe Bridge ≂ 1,970 ↔ 153
22 Onitsha Bridge ≂ 880 ↔ 1
23 Second Onitsha Niger Bridge ≂ 1,140

❹ Zambezi 2,574 km

1 T5 Bridge Mulemba ↔ 240
2 Cazombo Bridge ≂ 80 ↔ 502
3 King Lubosi Imwiko II Bridge ≂ 280 ↔ 246
4 Sioma Bridge ≂ 80 ↔ 118
5 Katima Mulilo Bridge ≂ 800 ↔ 146
6 Kazungula Bridge ≂ 500 ↔ 78
7 Victoria Falls Bridge ≂ 80 ↔ 418
8 Kariba Dam ↔ 63
9 Chirundu Bridge ≂ 315 ↔ 0.1
10 Sir Otto Beit Bridge ≂ 290 ↔ 577
11 Ponte Samora Machel ≂ 545 ↔ 5
12 Ponte Kassuende ≂ 700 ↔ 228
13 Dona Ana Railway Bridge ≂ 845 ↔ 63
14 Ponte Armando Guebuza ≂ 565

was the case with Schüßler-Plan in Sudan. The firms were on site as planners or construction supervisors for road and railway bridges and were involved in international teams or commissioned on their own. My impression from the trip has been reinforced. Bridge construction in Africa makes an important contribution to improving quality of life, and German civil engineering has a good reputation there. This is certainly also due to the fact that our colleagues deliver the quality expected of "Made in Germany" and also offer their local colleagues a genuine transfer of know-how. The flyer for our event is shown in the figure below Fig. 2. All speakers agreed to our request to contribute a statement on construction in Africa.

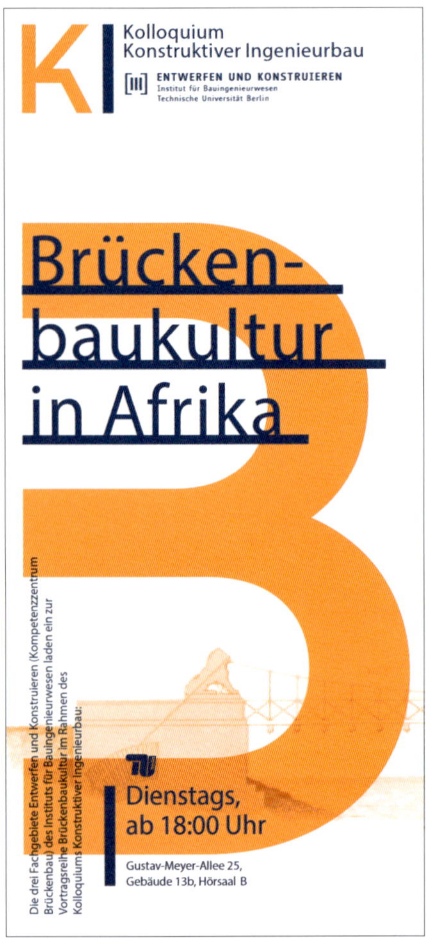

Fig. 2 Flyer for the lecture series 'Brückenbaukultur in Afrika' at the TU Berlin in the winter semester 2023/2024 (front)

It was only possible to find male speakers for the colloquium 'Brückenbaukultur in Afrika'. To compensate, only female civil engineers who report on *Brückenbaukultur* were invited to the next seminar in the winter semester 2024/2025.

Johannes Frühauf, SSF Engineers

'With the Standard Gauge Railway, an electrified railway line is being built in Tanzania in accordance with the latest European standards. As part of our projects, we were able to work with many well-qualified and motivated local engineers who will certainly continue to drive forward the development of civil engineering in the country. In the standard gauge railway project, we have learned how international cooperation can help to build knowledge and realise rail infrastructure that meets European standards.'

Gregor Gebert, DEGES (formerly with Schüßler-Plan)

'A chain of coincidences and the curiosity of the then Berlin-based managing directors of Schüßler-Plan led us to embark on our "African adventure". We were rewarded with insights into other cultures and – two new bridges over the Nile. Sudan has never had its own bridge-building culture. The bridges are functional in design and reflect the construction technology of their time. The first bridges were built by English companies, followed by Italian and later Chinese construction companies. We are proud to have been able to make a German mark together with a committed and competent Turkish construction company. The Nile bridges are an important chapter in my professional life as a bridge engineer.'

Martin Grassl, GRASSL Engineering Office

'Planning bridges from Germany is also characterised by the long distances involved. Short-term personal coordination is not so easy. However, digitalisation helps us a lot in this respect. But even more important is trust. This includes, in particular, the local client's trust in our work. We greatly appreciated this, and it was the guarantee of the project's success.'

Uwe Heiland, SEH Engineering

'Building and planning in Africa, or more precisely in Egypt, is clearly influenced by personal proximity and personal acceptance. It is pleasant to note that cultural differences with European partners are sometimes large, but differences in the approach to technical solutions and in the level of technical skills are sometimes small.'

Konstantin Papanikolaou, self-employed

'For me, Africa is a continent of unspoilt nature, fascinating wildlife, local customs and traditions, and – as I have experienced – cheerful people. Building a structure in Africa and ensuring that it is always in harmony with the above is the greatest challenge and, at the same time, an obligation that every planner must fulfil. Together with our African colleagues, we have managed to achieve these goals.'

Klaus Richter, INROS LACKNER

'For us, bridges in Africa, as in Europe, are primarily projects that serve the population by connecting people, cities, and regions. They are arteries along which cultural and economic activities develop. Bridges and roads are crucial transport infrastructure for Africa's development, as they promote trade, improve access to markets, integrate communities, facilitate education and healthcare, attract tourism and investment, and enable social progress. In short, they are the foundation for economic growth and human development on the continent. The projects are usually put out to tender in a public application process, followed by a transparent bidding process.

Infrastructure projects in Africa must meet the needs of the local population and comply with international standards and best practices. African authorities and companies should play the leading role in the planning and implementation of infrastructure projects in Africa, as they best understand local needs, circumstances, and priorities. This includes taking cultural, social, economic, and environmental aspects into account. That is why INROS LACKNER decided several years ago to implement projects in Africa primarily through its own local offices and with local employees. However, European experts and global expertise continue to provide valuable support in many areas. Through partnership-based cooperation that benefits from the experience and expertise of both sides, infrastructure projects can be made more effective and sustainable.'

Jörn Seitz, GAUFF Engineering

'The bridge-building culture in Africa is as diverse and multifaceted as the 54 countries on this continent. Bridges play an extremely important role in improving both the supraregional and local infrastructure for this continent of the future. There is considerable ground to make up, as is evident in the UN's master plan for north-south and east-west continental connections, such as the east-west connection through the swamps in Nigeria (Lagos-Calabar Coastal Road Project). The large and wide rivers of sub-Saharan Africa and West Africa pose a challenge for engineers and construction companies – but first and foremost, it is a question of financing. Funds for large-scale bridge construction can only be obtained through international financing, which is why companies from Asia or Europe are working with local construction companies to build these structures.

In planning and construction, increasing emphasis is also being placed on sustainability and environmental protection in connection with financing. This includes protecting ecosystems, minimising environmental impact, and taking climate change into account to ensure that bridges are durable in the long term and respect the environment.

Bridges are among the structures with which those responsible in the respective country also want to create a monument for the future, and so it is often the desire to choose special bridge architecture.'

Roads and Railways – Not Yet Connected

Anyone travelling by car in Africa quickly realises how important paved and safe roads are and how they cannot be taken for granted. Some 80 to 90 per cent of the transport of people and goods in Africa takes place by road [1], but 75 per cent of Africa's roads are unpaved and often impassable during the rainy season. More than half of the rural population has no access to roads that are passable all year round. [2] These statements are based on the World Bank's Rural Access Indicator (RAI): "proportion of the rural population who live within 2 km of an all-season road." [3] Without good access to safe transport routes, it is difficult to send children to school or get to a doctor. Agricultural products cannot reach cities or larger markets. Dependence on subsistence farming increases. There is often a lack of money or efficient administration to maintain existing paved roads and repair damage caused by overloaded trucks or flooding during the rainy season in a timely manner Figs. 1–2.

In Zambia, the Great North Road, National Highway 2, leads north to the Copperbelt and towards the Congo (DRC). Trucks coming from the north and loaded with heavy copper plates drive close together. They have carved decimetre-deep ruts into the soft asphalt, and next to the road we see cars that have skidded off the road as a result.

Fortunately, we are travelling in southern Angola during the dry season. The potholes on the EN140 to the Namibian border are so deep and large in places that we have to drive into and out of them at walking pace to get through safely. When the dust turns to mud in the rainy season, our car would probably have no chance here.

We encounter mud in Kenya near Sega, close to the border with Uganda. It has been raining, and the previously reliable reddish-brown earth of the village road has become so slippery that our car loses all traction despite its all-terrain tyres and driving at a near standstill. We come to a halt just before the deep ditch.

For us as tourists, these examples from Zambia, Angola, and Kenya are anecdotes, but for the population, they represent everyday suffering. We have often seen how locals try to make a virtue out of necessity. Market women and children use construction sites and road damage to offer goods, food, and drinks to slowed-down traffic. In his short story *A Hole in Onitsha* [4], Ryszard Kapuscinski entertainingly describes how a huge pothole filled with water makes it impossible to continue driving and forms "a dynamic, bustling and noisy neighbourhood" where stranded drivers are provided with food and accommodation. The problem remains that roads in Africa are not good everywhere.

Unfortunately, this is also reflected in the accident statistics. In 2014, 225,000 people died on Africa's roads, accounting for 25 per cent of global road deaths, even though there are far fewer vehicles on the roads there than in other regions of the world. [2] This high number is largely due to speed among motorcyclists. The authors of a 2020 report on road safety in Africa already estimate the number of annual road deaths at 296,000. [5] Most road users in Africa travel on foot, as passengers in buses or taxis. Based on surveys, the authors conclude that traffic fatalities are more frequent than in Europe because seat belts are not fastened, motorcyclists do not use helmets, and drivers are more likely to use their phones while driving Fig. 2.

There is also good news. I read: "Ethiopia, for example, has reduced the average distance to an all-weather road from 21 kilometres in 1997 to 12.4 kilometres in 2012, and access to an all-weather road is said to have decreased poverty by 6.9 per cent and increased food consumption by nearly 17 per cent." [2]

In Rwanda, I observed Chinese workers building roads as early as 1983. Today, the roads in this country are among the best I have seen. The paved main roads are well maintained, and where they pass through villages, there are pedestrian and bicycle paths separated from road traffic. Along the new road between Tunduma and Sumbawanga in eastern Tanzania, funded by the US, pedestrians are also safely separated from road traffic by kerbs.

In his lecture at the Technische Universität Berlin (→ Considerations – Bridge Building), Klaus Richter reported on the Bandajuma–Liberia Road and Bridges Rehabilitation Project. He supervised a 100-kilometre section and the construction of three bridges on this 400-kilometre route from Sierra Leone's capital Freetown to the border with Liberia. Before the new road that was partly funded by the EU was built, the 400 kilometres of unpaved road could be covered in three to five days during the dry season and in seven to fourteen days during the rainy season. Since a Senegalese construction company paved the road, this route can now be covered in six to seven hours! This is an impressive example of the efficiency of African road construction companies and the importance of road construction for the continent.

Road construction activities in Egypt are particularly striking (see also diary). As part of the Egyptian National Road Project, the military government under President al-Sisi built around 3,000 kilometres of roads between 2014 and 2024 [6], which accounts for around 10 per cent of the country's entire road network. This also includes the Egyptian sections of the Trans-African Highways TAH 1 from Cairo to Dakar and TAH 4 from Cairo to Cape Town. [7] At the same time, the National Authority for Tunnels has commissioned the construction of the sixth-longest high-speed rail network in the world. Siemens Mobility has also been involved in the construction of the 2,000-kilometre railway line, which will connect 60 cities in the country since 2021, with a contract worth a total of 8.1 billion euros. [8]

In South Africa, I find by far the largest and best road network on the continent. "South Africa has a road network three quarters of a million kilometres in extent, the tenth longest of any country in the world," write Malcom Mitchell and Brian Downie with pride in their richly illustrated, interesting, and well-researched book *Travels in the South – A History of South African Roads.* [9] One chapter is devoted to South African bridge construction. The authors link South Africa's first roads to Mapungubwe. Mapungubwe, South Africa's first indigenous kingdom, developed into the largest kingdom on the subcontinent between the tenth and fourteenth centuries. The roads were built for the gold and ivory trade with India and China.

In theory, it is now possible to cross Africa from south to north quite comfortably on the paved roads of the TAH 1. However, this journey is currently virtually impossible due to unrest and wars in Ethiopia and Sudan. Many other parts of the TAHs are still unfinished [7], and missing sections in Central Africa in particular make crossing from east to west cumbersome and slow down intra-African trade.

I found it difficult to obtain reliable, recent information on road construction in Africa. However, the diagram on the right shows that a country's prosperity is directly related to the availability of road and rail transport Fig. 1. This is where one should start and invest. The expansion of the Trans-African Highways TAH 1 to 10 [7], which have been planned since 1971 and cover a total length of around 56,000 kilometres, should be at the top of the continent's list of priorities, but as long as lawlessness and war prevail in the areas they pass through, implementation will probably remain a pipe dream Fig. 3. Does this also apply to the rail network in Africa? More on this in the next section on China.

Fig. 1 Correlation between prosperity and infrastructure

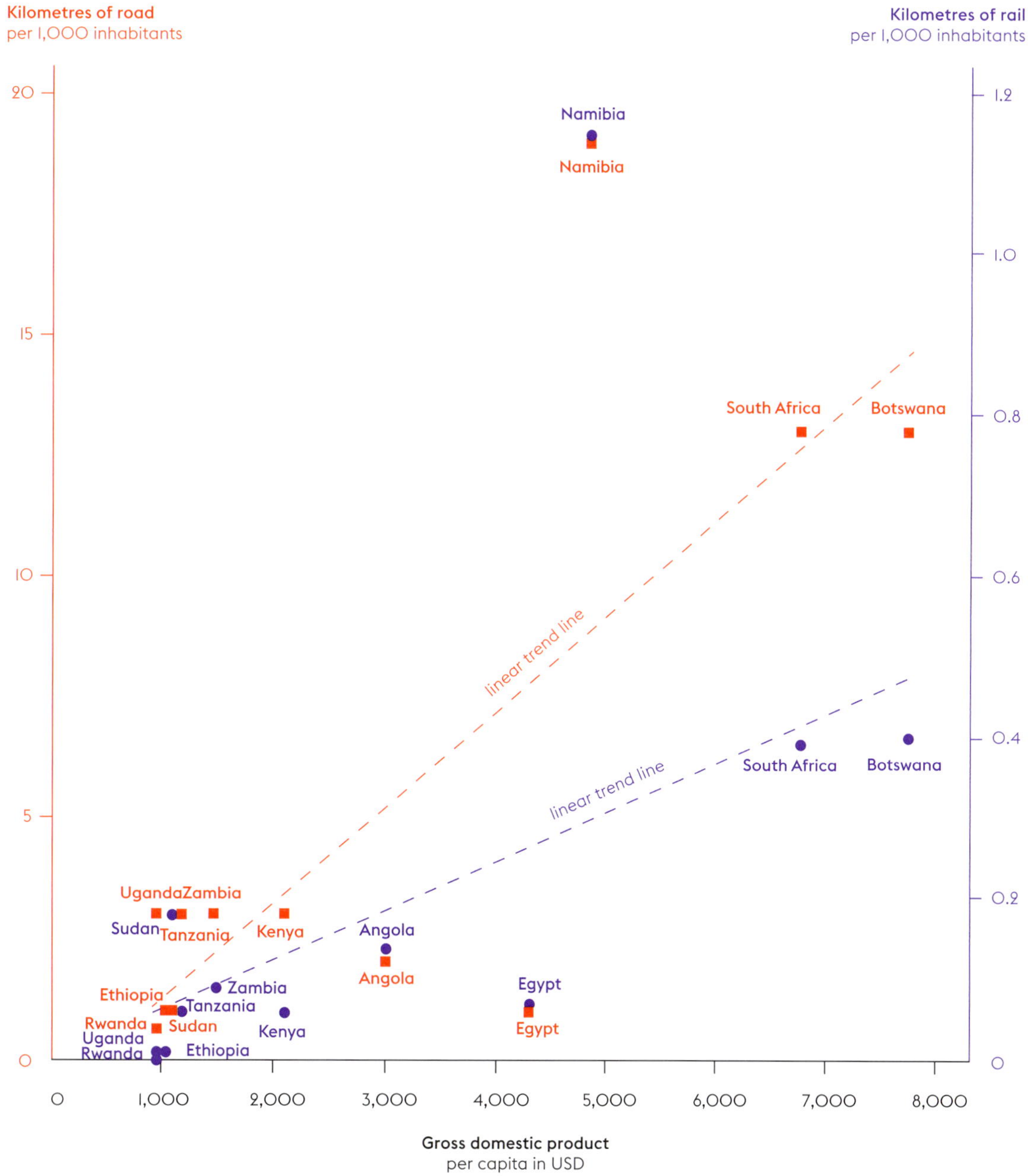

Source: World Bank | International Road Federation | CIA World Factbook | Wikipedia, see data sheet from page 220 onwards

Fig. 2 Expansion of road and rail networks compared to traffic fatalities

Source: World Bank | International Road Federation | CIA World Factbook | Wikipedia, see data sheet from page 220 onwards

Fig. 3 The Trans-African Highways

Source: Representation based on Rex Parry, Map of Trans-African Highways, Wikimedia Commons, 2007

China –
Bringing it into the Team

Why am I dedicating a chapter to China in this book? Anyone travelling through Africa encounters this country at every turn. Skyscrapers, hotels, football stadiums, bridges, roads, and railway lines, whether under construction or already in operation, vividly illustrate China's activities in the construction sector.

→ In Menongue, Angola, the shell of a massive multi-storey concrete block near the airport, next to the EN280 national road, is awaiting completion and inauguration as a multi-star hotel. We see this type of building, obviously of Chinese design, time and again in Africa Fig. 1.
→ The National Heroes Stadium in Lusaka, Zambia, completed in 2014, can seat 60,000 spectators and was built by the Shanghai Construction Company in three years. Driving past, I can see that it lacks maintenance. The lightweight steel truss cantilevers of the roof structure are visible under the torn membrane roof Fig. 2.
→ Near Kabingo in Tanzania, we struggle along a seemingly endless, reddish-brown dusty track. Next to us, the China Henan International Cooperation Group is already building the 260-kilometre-long Kabingo–Kasulu–Manyovu–Mugina Road, which will be asphalted in the future and will soon make fast and comfortable travel on this route possible Fig. 3.
→ In Kenya, heavy single-span bridges of Chinese design made of precast prestressed concrete elements carry the 600-kilometre Standard Gauge Railway (SGR) – coming from Mombasa – across Tsavo National Park. Several hundred kilometres further on near Suswa, these bridges, built by the China Bridge and Railway Corporation, reappear on their way to Naivasha Fig. 4.

The expansion of the building stock and infrastructure in Africa and the associated potential for the positive development of these countries cannot be overlooked. Is the impression of strong Chinese involvement in this area correct?

After Mao Zedong's death in 1976, during the economic opening of the country by his successor Deng Xiaoping, who ruled the vast country from 1979 to 1997, domestic issues were on China's agenda. Trade with Africa accounted for only about 1 per cent of China's foreign trade volume at the end of the 1980s. Africa's trade with China also accounted for less than 1 per cent of its total trade volume during this period. [1] Since then, first under Jiang Zemin, then under Hu Jintao, and now Xi Jinping, China's involvement in Africa in terms of development aid, lending, trade, and investment has increased significantly. Today, China is the most important trading partner for many African countries (Figs. 6–9).

1

2

3

4

5

Figs. 1–5 Chinese construction projects we encountered along the way: in Angola (1), Zambia (2), Tanzania (3), Kenya (4), and Ethiopia (5)

The China Africa Research Initiative (SAIS–CARI) at the Johns Hopkins University School of Advanced International Studies in Washington has been studying relations between Africa and China since 2014 and has collected a wealth of valuable data on construction and published some very interesting articles. [2] According to the National Bureau of Statistics of China, the gross annual revenue of Chinese construction companies in Africa increased from 1.9 billion US dollars to 37.1 billion US dollars between 1998 and 2021. This corresponds to 27.8 per cent of the foreign turnover of Chinese construction companies, putting Africa in second place after China's neighbouring countries in Asia (50 per cent). In 2018, Africa ranked fourth in terms of foreign revenue generated by the German construction industry, with 314 million euros and 7 per cent, behind Europe (63 per cent), Asia (19 per cent), and Latin America (10 per cent). [3] The market share of Chinese construction companies in Africa had already reached 62 per cent in 2018. At the time, the German Construction Industry Association complained that this high figure was due to "China's extremely generous lending to African countries in the infrastructure sector, which is linked to the condition of tying supplies to Chinese companies," and to the fact that European financing has shifted from Africa's infrastructure to education, health, and civil society. [3]

The five Chinese construction companies with the highest turnover in Africa in 2019 were [4]:

1. China State Construction Engineering Corporation, Ltd.
2. Huawei Technology Corporation, Ltd.
3. Sinohydro Corporation, Ltd.
4. China Civil Engineering Construction Corporation (CCECC)
5. China Road and Bridge Corporation (CRBC)

China's influence in Africa is also supported by a growing Chinese diaspora. In 2014, around one million Chinese immigrants were already living in Africa, opening shops and businesses, importing Chinese goods, and consuming and disseminating Chinese media content. Are they turning Africa into *China's Second Continent* [5]?

China's involvement in Africa has moral, political, and commercial reasons. In addition to providing assistance and gaining more influence, it needs contracts for its construction companies, new markets for its products, and secure access to resources in Africa. These are all good reasons to invest in road, rail, bridge, water, and building construction there. If these infrastructure measures benefited both the countries and the local population equally, it would create a win-win situation – but this is precisely where opinions differ.

Is China's influence a development opportunity or a debt trap? There is no doubt that it is an attractive offer to have dilapidated and inefficient infrastructure replaced in order to increase regional and national connectivity and to reach markets, schools, and hospitals more conveniently, cheaply, and quickly. However, this requires African countries to take out loans and be able to service them. I have heard several times from African colleagues that European lenders tend to finance smaller projects and tie their loans to compliance with standards relating to democracy, the rule of law, human rights, and corruption. Chinese lenders, on the other hand, also requires collateral for their loans, but otherwise apply a "no questions asked" policy. How does the resulting cooperation between China and Africa work?

In his book *Die Seidenstraßeninitiative in Ostafrika: Auf dem Weg zu einer Hegemonie Chinas?* [6] *(The Silk Road Initiative in East Africa: Towards Chinese Hegemony?)*, Simon Züfle explores this question using the example of various major projects in the Chinese Silk Road Initiative (SRI). He looks at the Chinese-financed railway line from the Ethiopian capital of Addis Ababa to Djibouti and the Kenyan railway line from Mombasa to Malaba. For his study, he evaluated hundreds of newspaper articles in both Ethiopia and Kenya.

The construction of the approximately 750-kilometre-long Standard Gauge Railway (SGR) from Addis Ababa to the port of Djibouti, which was inaugurated in 2017, has connected landlocked Ethiopia to the sea, promoted the country's development, created jobs, ensured the transfer of know-how, and brought more Chinese tourists to the country. The construction of the railway line, which cost just under 3.5 billion US dollars, anchored China in the Ethiopian economy and secured it follow-up contracts. However, because the Ethiopian power grid was initially unable to supply the electrified line with sufficient power, due to quality defects in the rail vehicles, and accidents, the line is far from being as efficient as hoped. This is perpetuating Ethiopia's national debt and has resulted in a loss of 1 billion US dollars for the Chinese insurer Sinosure. [6]

The 950-kilometre-long, non-electrified SGR railway line for passenger and freight transport in Kenya is intended to connect the port city of Mombasa on the Indian Ocean with Malaba on the border with Uganda. So far, the nearly 600-kilometre-long line from Mombasa to Naivasha, about 120 kilometres northwest of Nairobi, has been completed at a cost of approximately 5 billion US dollars. This project is also accompanied by high hopes for national development and for connecting the neighbouring countries of Uganda and Rwanda. However, accusations arose early on regarding the non-transparent awarding of contracts, corruption in Kenyan authorities, quality defects in the rolling stock, and environmental damage in the national parks that the line crosses. Kenya's largest infrastructure project since independence caused the country's debt to skyrocket. The loans are difficult to service because – partly due to an overly optimistic Chinese feasibility study – the line is only operating at 50 per cent capacity. [6]

Another complicating factor is the fact that the construction of the railway line by CRBC and the compensation of landowners (by the Kenyan government) were decoupled, and compensation was not paid or was not paid on time. "This model of labour division is believed by many Chinese companies to be the most cost-effective way to operate in African countries. However, to Chinese companies' chagrin, this liaison method can be regarded by the layperson as dangerous and corruptive, giving rise to conspiracy theories and mistrust among host societies." [7]

Extensive collateral for the loans shows that in the event of payment default, China does not directly seek subsequent ownership of the investment, such as the port of Mombasa in the case of the railway line described above. [8] But even unplanned debt traps can snap shut. In November 2020, Zambia became the first African country to default on its outstanding obligations. [6] For reasons that lie on both sides, however, large-scale projects such as those described above have not proved to be successful tests of the African-Chinese combination of development cooperation and economic interests. Perhaps this is one of the reasons why the turnover of Chinese construction companies in Africa has declined considerably since 2015. China appears to have become more cautious with construction investments, as can be seen from the graph based on data from CARI Fig. 10. Simon Züfle confirms: "The decline in the construction boom is due to domestic political reasons in China (questioning of high investments abroad, economic stagnation/crisis), the slow repayment of outstanding loans by African governments, and a growing scepticism of African governments towards China."

This is worrying because I believe that Africa will need to cooperate with China in order to tackle the tasks ahead. China is the only country in recent decades that has mastered tasks on a scale similar to what Africa is facing. China has gained experience with large-scale infrastructure projects in its own country that hardly any other country in the world can offer: "Scope, scale, and speed are qualities that China brings in ways that others cannot." [7] The countries of Africa have, nolens volens, had much longer and more intense experiences with the former colonial powers of Europe than with China. Today, these experiences have led to established cooperation in the fields of education, health, and civil society, as well as to far-reaching reassessments of colonial injustice. Attempts at reconciliation, financial reparations, and serious restitution debates [9] are underway. Africa could take advantage of the good intentions of the former colonial powers. If Africa also succeeds in involving the (digital) power of the USA in the cooperation, a global alliance for its urgently needed progress would be within reach. Is it a naïve illusion to believe that Africa will be able to use the experience and opportunities of the powerful trio of China, Europe, and the USA for its future? Perhaps, but in my opinion there is no alternative.

Fig. 10 Annual turnover of Chinese construction companies in Africa

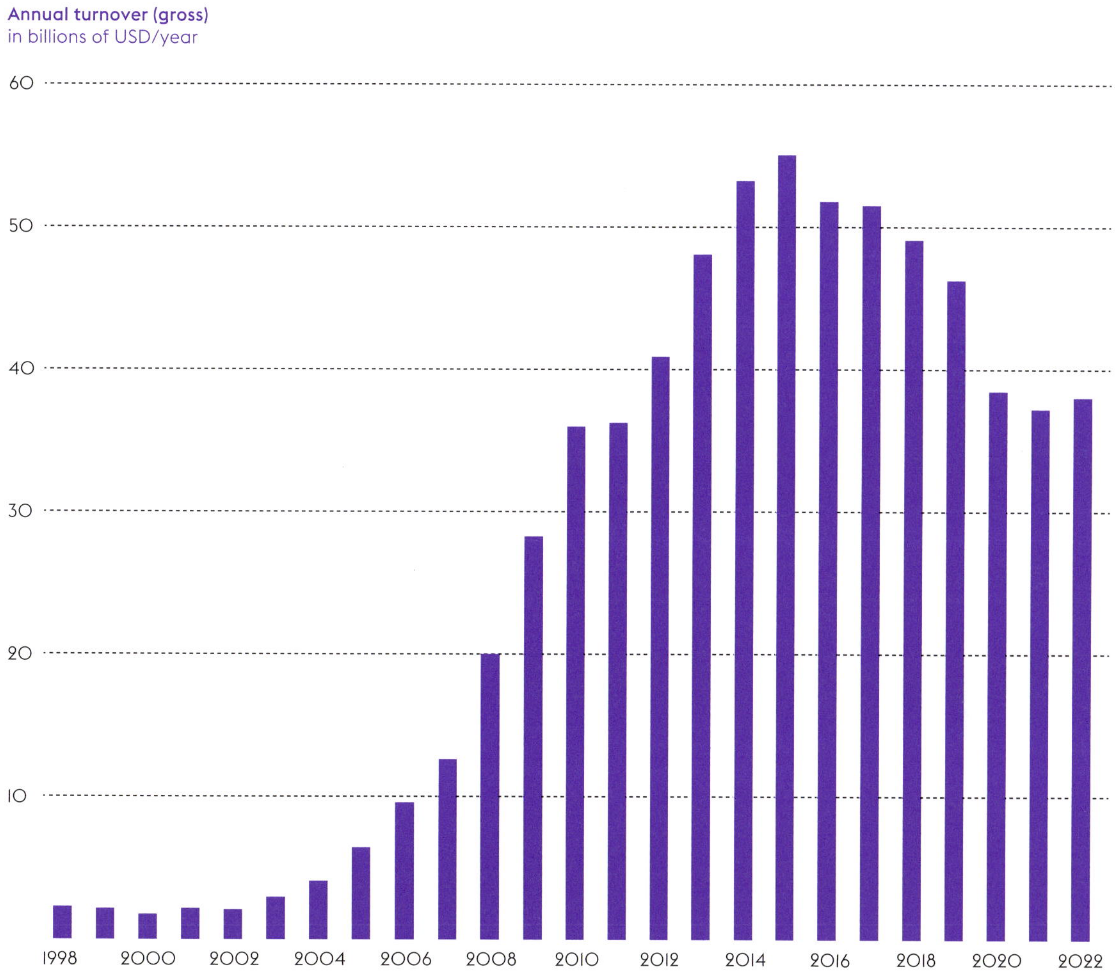

Source: Representation based on Johns Hopkins University, SAIS-CARI, 2024 [2]

Fig. 6 Trade flows in Africa: Top 3 import and export partners: **Total economy**

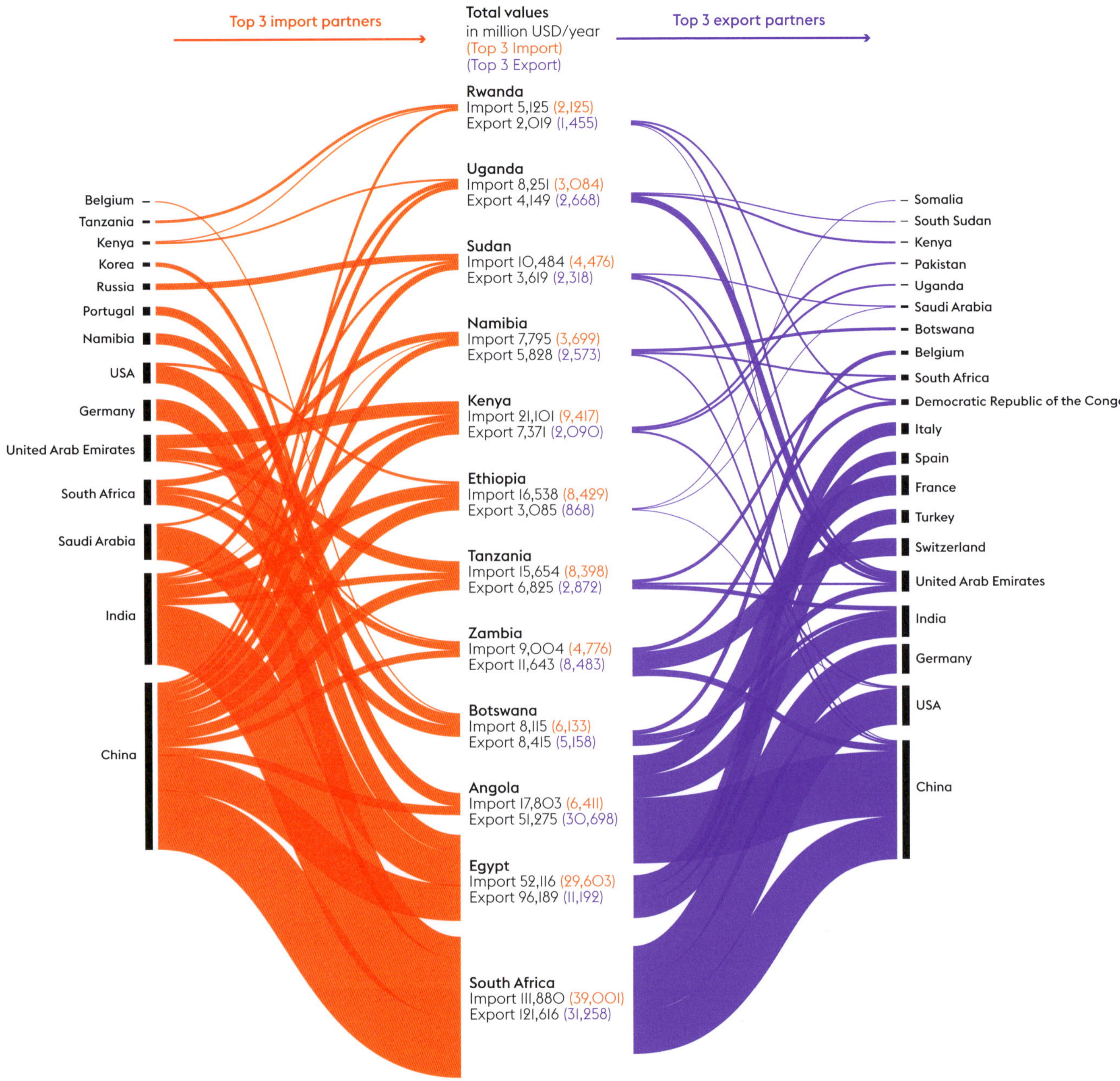

Source: UN Comtrade Database, see data sheet from page 220 onwards

Fig. 7 Trade flows in Africa: Top 3 import and export partners: **Cement & concrete**

Source: UN Comtrade Database, see data sheet from page 220 onwards

Fig. 8 Trade flows in Africa: Top 3 import and export partners: **Iron & steel**

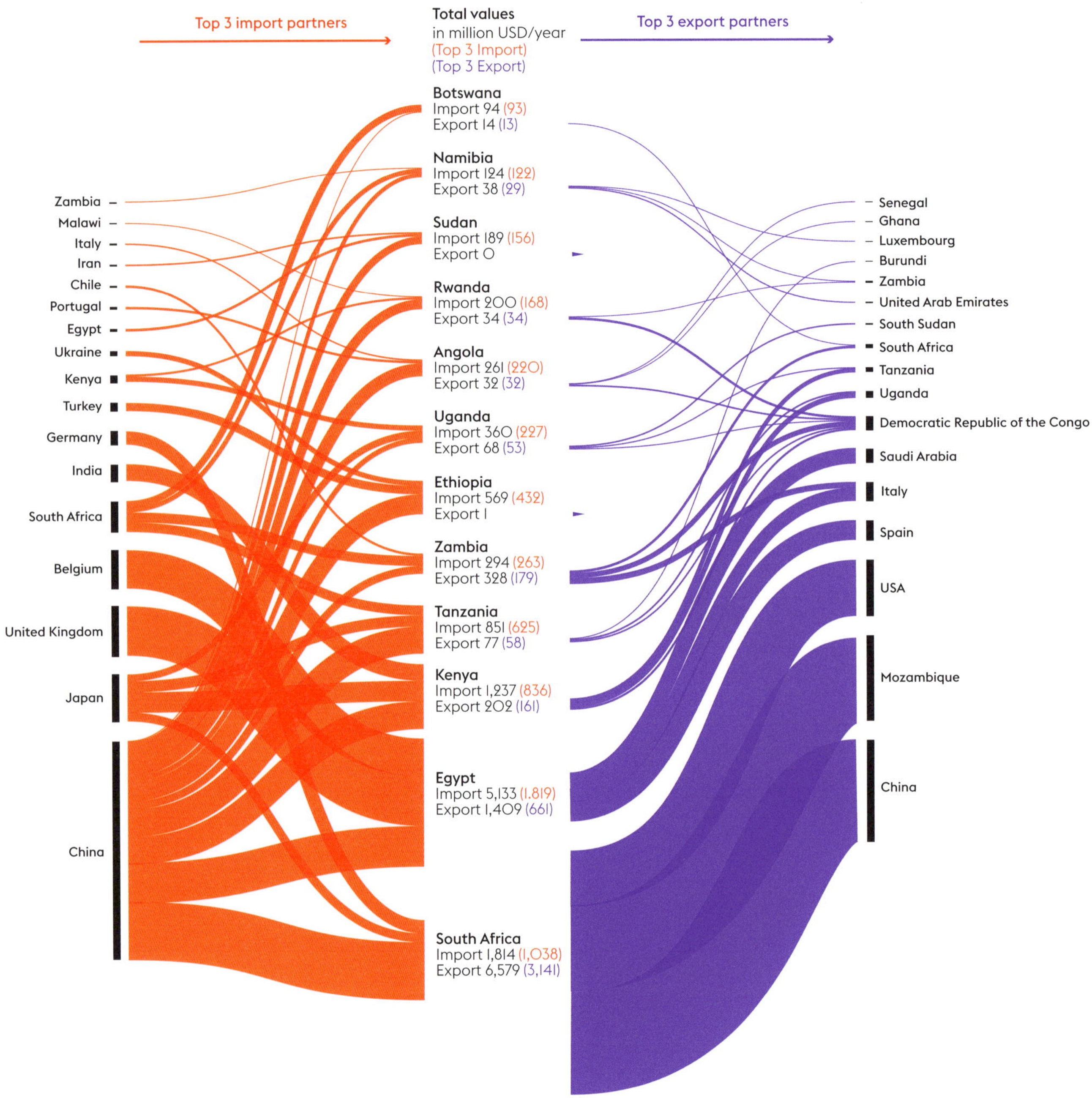

No significant amounts
Source: UN Comtrade Database, see data sheet from page 220 onwards

Fig. 9 Trade flows in Africa: Top 3 import and export partners: **Raw timber**

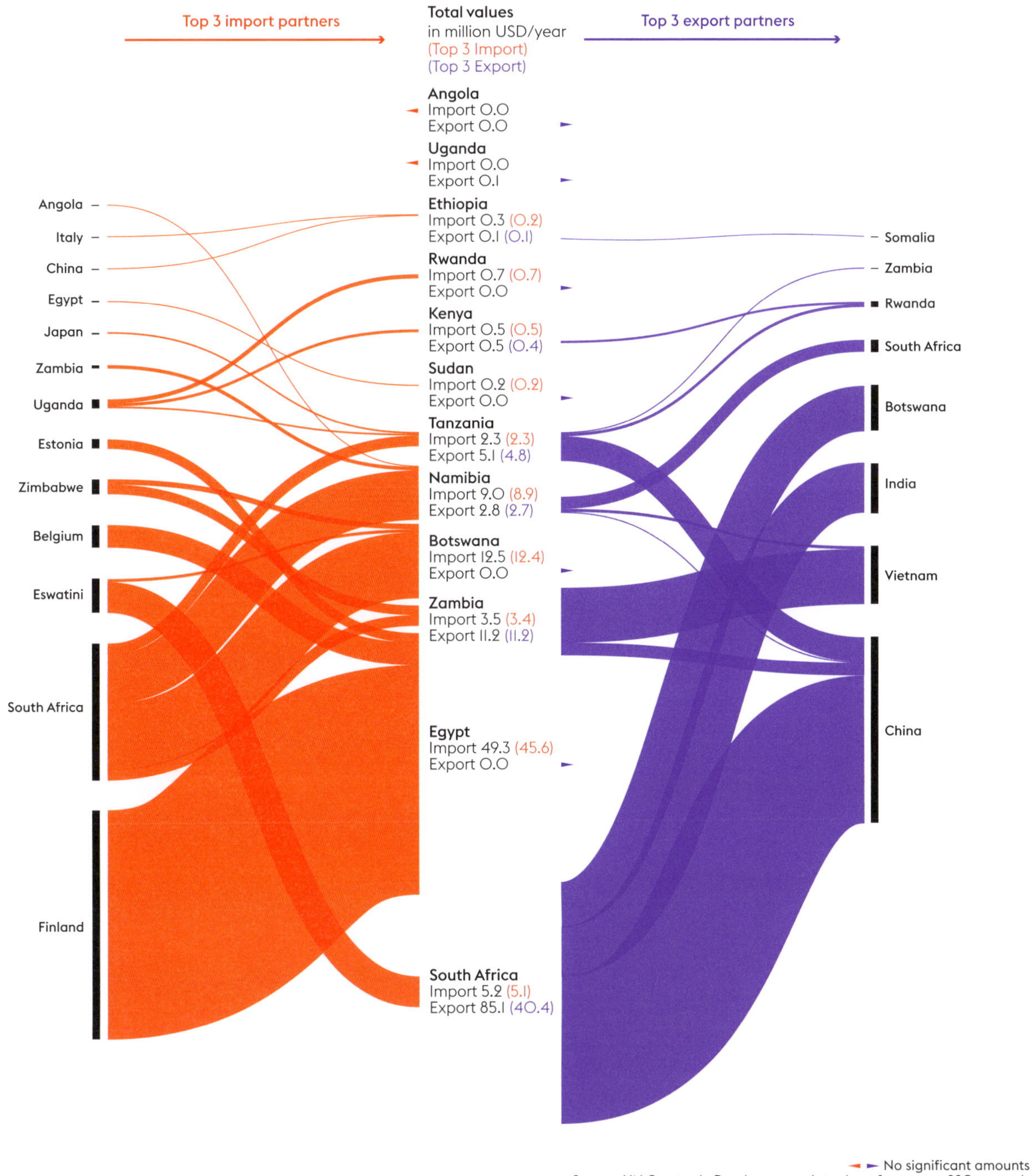

Source: UN Comtrade Database, see data sheet from page 220 onwards

Residential Construction – It's High Time

Just as a bridge is the prototype of an engineering structure, a house is a representative example of built architecture. Perhaps, as a civil engineer, I find it difficult to relate to residential construction because I am conditioned by my profession to think in technical terms. A bridge is primarily created from the technical local context, often characterised by difficult terrain, high loads, and complex construction methods. Of course, cultural and formal aspects are important, and technical solutions are also required in residential construction. But social issues play a much greater role there. How do I embed the house in its urban or rural environment, how do I orient it towards the sun and protect it from the weather, where do I need light, which surfaces are appealing, what is the internal layout, where do I store things and where do I cook and sleep? The supporting structure must be safe and economical, but compared to a bridge, which is almost entirely a supporting structure, it plays a subordinate role in a house.

In Africa, an incredible amount of new living space will have to be made available in the coming years and decades. Between 2020 and 2050, the population is expected to grow by around 1.2 billion people. [1] Rural exodus has already given rise to megacities and conurbations such as Cairo, Kinshasa, and Lagos, with more than 15 million inhabitants. [2] During my trip, I heard time and again: "We need decent and affordable housing. This is the main challenge for engineers and architects in our country." Providing housing is undoubtedly the order of the day, although "decent and affordable" is defined differently from country to country and depending on the population group. For people living in slums, this means something completely different than for members of the so-called middle class, who have very different financial resources at their disposal. 'Decent and affordable' is understood differently in the city than in rural areas.

During our trip, we see large refugee camps in the conflict areas. Where aid organisations cannot set up tents, people live in self-built shacks of every conceivable design. In the cities, the informal settlements that are springing up everywhere are impossible to overlook. We avoid venturing into them, but even from a distance, the cheap corrugated iron huts, rubbish dumps, and open sewers show that people cannot live there in dignity. At the same time, great efforts to provide housing are visible everywhere, and settlements with new single-family homes are also part of the picture Figs. 1–6.

Residential construction in Africa was the subject of the exhibition *Afritecture – Bauen mit der Gemeinschaft (Building with the Community)*, which was shown at the Architecture Museum of the Technische Universität Munich in 2013. The accompanying book *Afritecture* [3] provides a nice overview of the multitude of architectural approaches and locally anchored reference buildings that demonstrate ecological and social sustainability. Andres Lepik, the editor, says: "One of the central tasks for planning in the coming years will remain the further growth of the city and, in particular, that of informal settlements." My thoughts on this complex and multi-layered topic are also informed by discussions with three experienced experts. They are well acquainted with residential construction in Africa and Europe and take a holistic view of it.

German civil engineer Simone Stürwald is a professor at the Institute for Construction and Environment at the University of Applied Sciences of Eastern Switzerland in Rapperswil and has been working on issues relating to residential construction in Africa for some time. In the summer of 2022, she spent a research semester in Africa – what a coincidence. [4] Over the years, she has built up a strong network in Africa and has already got to know many universities, NGOs, companies, and representatives of the financial sector there. During her research semester, she visited the cities of Dar es Salaam, Lagos, Cairo, Kampala, and Johannesburg, partly to familiarise herself with traditional construction methods and sustainable building using local resources.

She reports on manioc shell ash or volcanic ash, which can be added to concrete to save cement, and on reuse techniques used by those with fewer financial resources, who skilfully reuse discarded building materials for their homes. She cites the fact that bank financing is practically impossible for the poorer segments of the population as an obstacle to the upgrading of slums. In many countries, she has also noticed the problem of unclear ownership structures and vaguely defined building zones, which make construction difficult for local people.

Simone Stürwald's conclusion on housing construction in Africa: "The push to do something should come from Africa, and in my opinion, the solutions for housing construction in Africa should also be developed there. There are now excellent civil engineers and architects there. We can engage in technical dialogue and learn from each other, but the impetus must come from them and the responsibility must remain with them."

Architect Fabienne Hoelzel, a Swiss citizen and German national, is a professor of design and urban planning at the State Academy of Fine Arts in Stuttgart, Germany. She is intensively involved with the infrastructure of informal settlements in large cities such as Lagos. Her finding is that services such as water supply and sewage, transport of people and goods, and waste disposal function quite well there despite the lack of traditional urban service providers: "but in a different way and, above all, thanks to the work of local women." [5] She not only

1

2

3

4

5

6

Figs. 1–6 Single-family housing estates in South Africa (1), Namibia (2), Tanzania (3), and Ethiopia (4). In contrast, an informal settlement in Nairobi, Kenya (5). And those who can afford it build with columns ... (6)

researches the interrelationships of informal infrastructure, she also works in the surely not hazard-free settlements of Lagos. Together with local partners and the local population, she develops strategies for more sustainable houses with better supplies and hygiene. One result is the *Building Guide for a Flood Resilient Waterfront Community* in Lagos. [6] She has been active in the poor neighbourhoods of this megacity for over ten years, and in 2013 she founded the architectural firm Fabulous Urban with an "activist design approach" [7].

Fabienne Hoelzel does not think much of official requirements to produce as many 'units' as possible. Many Africans cannot afford any of the houses offered by government programmes. Where she travels, even 1,000 US dollars is often too much for a 'house'. She believes it is more important to provide the local population, who are relatively well equipped to build themselves, with simple infrastructure and manuals with tips for improvement. She also considers the land and property issues that often remain unresolved to be a major obstacle to sustainable construction, and she laments the lack of mortgage models and the many houses that stand empty as a result of corruption.

Architect Luyanda Mpahlwa comes from South Africa. During apartheid, he had to spend five years in the notorious Robben Island Prison off Cape Town because of his political views before he was able to leave for Germany with the help of Amnesty International. Once in Germany he studied architecture at the Technische Universität Berlin, among others with Rainer Hascher and Peter Herrle, whom he describes as having had a formative influence on his education. Since returning to South Africa, he has explored both the theory and practice of 'socially just and relevant' architecture in his home country and on the continent: "A theory of African architecture will be less about buildings and more about cities and spacemaking in order to suit the African way of life." [8]

In a video call conducted mainly in German – a picture of Nelson Mandela hangs on the wall behind Luyanda Mpahlwa – he first highlights the advantages of traditional local residential construction: "Natural building materials, such as clay for the walls and straw for the roofs, ensure a good indoor climate and good thermal properties. With their small windows and no mechanical ventilation, they are exemplary examples of climate-friendly construction. This type of construction requires local knowledge, and the buildings can be constructed inexpensively with local labour and local materials. Because of today's preference for working with industrially manufactured and often imported building materials, there is a risk that such construction practices, experience, and thus sustainability in construction will be lost. In the area where I come from, people used to build their own houses. There is no word for 'architect' there. The question is: how can we architects apply this knowledge today?"

One of his most important projects is the Informal Settlement Kosovo in Philippi Township near Cape Town, where he was commissioned by the Department of Human Settlements in 2016 to plan "Informal Settlement Upgrading" and 5,000 apartments for the 15,000 residents. There are no roads or street lighting, only illegal electricity connections, no running water, no sewage system, no waste disposal, and too few toilets. There are no social facilities, only tin shacks that the people there have built themselves. "A real slum settlement" that has existed since 1999. Most of the residents are unemployed and crime is high; it is a dangerous area. During his visits, he was surprised at how well the residents know each other and how well they organise themselves. An active "community network" and also an economic network has developed. The informal sector is booming and "land distribution is carried out illegally on site". There are informal nurseries because mothers work, and a priest runs a self-built church that serves as a clinic at night until the next morning. No ambulances, police, or fire brigades come here after dark. For Luyanda Mpahlwa, Kosovo is representative of many informal settlements on the outskirts of Africa's major cities.

In South Africa, he believes, a clear distinction must be made between building in rural areas and in cities. Western models for building according to standards and regulations can be applied to cities, and may even be necessary there, but they mean that it takes a very long time from the decision to build to the planning and implementation stages. In his Kosovo project, he has only built 400 houses in eight years. Allocating building land is difficult, and the negotiation and approval processes are complex. Mpahlwa feels trapped in a vicious circle while the need for housing is growing. He says I have to imagine building in rural areas as being completely different, as society here is still anchored in traditional structures. The land is managed as communal property by local chiefs and kings. "The chiefs are the custodians of ancestral land, and this has been common practice since prehistoric times." Private property is rare, decisions are made quickly, and construction is often a joint effort. This reality contrasts with, and perhaps conflicts with, the guidelines and processes of urban settlement planning. Mpahlwa wonders whether standards and architects are even necessary in rural areas. Anyone who wants to build there will fail if they do not engage with the existing structures and negotiation processes. He is convinced that projects coming 'from outside', without knowledge transfer and only with imported labour and materials, have no chance in Africa.

Luyanda Mpahlwa's assessment of his country: South Africa is burdened by the legacy of 'apartheid spatial planning'. The housing conditions remain highly unequal between rich and poor and between black and white people. The townships developed for the black population are located on the outskirts of cities, are overcrowded, and have no connection to the sewage system. Informal settlements have multiplied since 1994. A new housing policy addressing these inequalities and inhumane living conditions is urgently needed. According to Mpahlwa, the biggest problem is that government programmes respond to the housing shortage with quantitative targets, i.e. to produce as many 'units' as possible. The quality of the settlements is not a priority, he says. Politically speaking, this is understandable and perhaps necessary because the new government wants to fulfil its social duty and care for historically excluded population groups. But wouldn't it be more important to ask: Who are we building for? What do residents want, and what do they really need? How can we distinguish which housing models are suitable for the various groups affected? At the moment, the focus is more on finding "one solution for all", regardless of whether in urban or rural areas. This creates a gap between the goals of government programmes and reality, and the small projects and experiments of NGOs are often not implemented or taken seriously. There is still no official, well-thought-out strategy that can be implemented to deal with informal settlements, "no conceptual thinking on alternative models", and too little research into solving housing problems, he says.

Luyanda Mpahlwa cites historical examples in Germany, where at the beginning of the twentieth century, the provision of affordable housing for the working class was a major issue. For him, the idea of building complexes with commercial courtyards, such as the Hackesche Höfe in Berlin, and other programmes of that time – like the Bauhaus with its early approaches to social housing and the Frankfurt Kitchen – represent ways to improve people's quality of life. Despite different constraints, he also sees successes like this in Africa. South Africa needs an integrated approach in which politics, civil society, and the various planning disciplines work together. "We need dialogue towards a new urban agenda!" He is hopeful. At the end of our conversation, he refers to promising initiatives he collaborates with such as the African Centre for Cities (ACC) at the University of Cape Town (UCT), which deals with 'Urbanism from an African perspective'. [9]

Simone Stürwald, Fabienne Hoelzel, and Luyanda Mpahlwa share the view that holistic political concepts for residential construction in Africa are still largely lacking; that ethnic and cultural aspects are decisive for acceptance and success; and that impetus for good construction can and must come 'from within'.

Germany also lacks affordable housing, with complaints about too few social housing units, too much bureaucracy, and inconsistent policies. In Africa, however, especially in the Sahel region where the population is growing most rapidly, the scale of the problem is entirely different. As shown above, there is no shortage of approaches and exemplary projects. However, I believe that we civil engineers can become even more involved in interdisciplinary cooperation, even if residential construction does not appear to be an original civil engineering topic.

This is misleading. Many of the considerations raised in this book can be found in this field. Civil engineering education at African universities needs to impart knowledge and raise awareness of resource-efficient construction in the local context. As in Germany, this requires professors with practical experience. Solar energy not only makes it possible to switch to heating, cooling, and cooking without wood, but also to produce building materials with lower emissions. Africa can leapfrog the fossil fuel era. In addition to traditional self-build construction, concrete is unavoidable, but we must use it sensibly and try to obtain it from natural regional sources. This will enable inexpensive, durable, and sustainable construction for many people. Rural exodus will only decline when roads and bridges that can be used all year round provide the rural population with access to schools, markets, and medical care. Can't Africa harness the power of global cooperation, including that of China, to achieve this? Is residential construction in Africa really a 'mission impossible'? What happens in Africa affects us all. That is why we civil engineers should get involved in construction in Africa, together with all stakeholders, without prejudice and with our professional expertise.

Concluding Remarks

Since returning from my journey through 12 African countries, research and reflection have supplemented the rather fragmented knowledge I gained in 2022 and my memories of the journey so that it can be better understood as a whole. The highlights of the journey have been recorded and may one day allow for comparison from the perspective of a later present.

For example, I learned how many civil engineering contacts have already been established between Germany and African countries and how many joint projects already exist. I couldn't mention them all in this book, but I am delighted that I am now connected with numerous experts on the subject of 'building in Africa'. Colleagues from Egypt, Ethiopia, Namibia, and South Africa have since visited me in Berlin, conducting research or working with us at the Technische Universität Berlin and at our consultancy (schlaich bergermann partner). These joint ventures are continuing, and my next visits to conferences and construction sites in Africa are coming up. In a series of lectures at the TU Berlin, German civil engineers reported to students and guests from the industry about their projects in Africa, and 12 students devoted their bachelor's theses to 'building in Africa'.

Some topics will continue to occupy my thoughts. As everywhere else in the world, social and economic developments are leading to conflicts in Africa, though to a more extreme degree: How will the ever-growing population and the increasingly threatened 'nature' be able to coexist there? Hopefully, no false dichotomies will be created, pitting people against nature and nature against people. But what does good coexistence look like? During my trip, I developed a growing sense of shared responsibility and an urge to take action. At the same time, I have to accept the visible, but also the missing momentum on the ground as it is. And I feel the same way about African civil engineering culture: it cannot be understood primarily from the perspective of industrialised nations, nor solely from a comparative perspective. Although Africa is already present in the global history of building with its own independent development, this does not yet apply to civil engineering history. This is particularly true given that global civil engineering history itself has considerable gaps – and that the canonisation process will continue to evolve.

Many questions remain unanswered, but at least I hope that this 'engineering safari' will be followed by further initiatives and projects. The journey is not over yet.

Data Sheet

Thematic area	Indicator	Unit	Ref. year	World	Germany	South Africa	Botswana	Namibia	Angola
Society	Population 2022	[mil. inhabitants]	2022	7,951	83.8	59.9	2.6	2.6	35.6
	Population 2050 (estimated)	[mil. inhabitants]	2022	9,668	78.6	73.5	3.7	3.8	72.3
	Population change 2022–2050	[change in %]		+22	-6	+23	+42	+46	+103
	Population change 2022–2050	[change in mil. inhabitants]		+1,717	-5.2	+13.6	+1.1	+1.2	+36.7
	Annual population change	[%]	2022	+0.8	+0.7	+0.8	+1.6	+1.5	+3.1
	Land area	[km²]	2022	-	357,022	1,219,909	582,000	824,292	1,246,700
	Population density	[inhabitants/ km²]	2022	-	234.7	49.1	4.5	3.2	28.6
	Average age	[years]	2022	30.2	45.0	27.9	22.6	20.8	16.4
	Human Development Index (HDI)		2021	-	0.95 (7/193)	0.72 (110/193)	0.71 (114/193)	0.61 (142/193)	0.59 (150/193)
	Inequality-adjusted HDI (IHDI)		2021	-	0.88	0.46	0.49	0.40	0.34
	Overall loss due to Inequalities (IHDI/HDI – 1)	[%]	2021	-	-7	-36	-31	-35	-42
Economy	GDP per capita	[USD]	2022	12,730	48,718	6,767	7,726	4,896	2,934
	Economic growth, annual change in GDP	[%]	2022	3.1	1.8	1.9	5.5	5.3	3.0
	Remittances	[% of GDP]	2022	0.8	0.5	0.2	0.3	0.4	0.0
	Unemployment rate	[%]	2022	5,8	3.0	29.8	20.7	20.8	10.2
	Inflation rate	[%]	2022	8.0	6.9	7.0	11.7	6.1	21.4
	Trade balance	[% of GDP]	2022	0.6	2.0	2.0	1.5	-18.4	19.1
	Public depth, relative government depth	[% of GDP]	2022	-	66.1	71.1	17.8	70.5	64.8
Politics	Form of government		2023		Federal parliamentary republic	Parliamentary republic	Parliamentary republic	Presidential republic	Presidential republic
	Year of independence				-	1931[1]	1966	1990	1975

Zambia	Tanzania	Rwanda	Uganda	Kenya	Ethiopia	Sudan	Egypt	Source/*formula*/comment
20.0	65.5	13.8	47.2	54.0	123.4	46.9	111.0	World Bank (https://data.worldbank.org/indicator/SP.POP.TOTL)
37.5	129.9	23.0	87.6	85.2	214.8	84.5	160.3	World Bank (https://databank.worldbank.org/source/population-estimates-and-projections)
+88	+98	+67	+86	+58	+74	+80	+44	*(Population 2050 / Population 2022) – 1*
+17.5	+64.4	+9.2	+40.4	+31.2	+91.4	+37.6	+49.3	*Population 2050 – Population 2022*
+2.8	+3.0	+2.3	+3.0	+1.9	+2.5	+2.6	+1.6	World Bank (https://data.worldbank.org/indicator/SP.POP.GROW?locations=CM%26name_desc=false)
752,612	947,303	26,338	241,550	580,367	1,104,300	1,861,484	1,001,450	Wikipedia (https://en.wikipedia.org/wiki/List_of_countries_and_dependencies_by_area)
26.6	69.1	524.0	195.4	93.0	111.7	25.2	110.8	*Population 2022 / Land area*
17.3	17.1	19.2	16.3	19.3	18.5	18.1	23.8	Our World in Data (https://ourworldindata.org/grapher/median-age?tab=table&time=2022)
0.57 (153/193)	0.53 (167/193)	0.55 (161/193)	0.55 (159/193)	0.60 (146/193)	0.49 (176/193)	0.52 (170/193)	0.73 (105/193)	United Nations Development Programme (https://hdr.undp.org/data-center/specific-country-data#/countries)
0.34	0.37	0.38	0.38	0.44	0.32	0.33	0.56	United Nations Development Programme (https://hdr.undp.org/data-center/specific-country-data#/countries)
-40	-30	-31	-32	-27	-34	-36	-23	*IHDI / HDI – 1*
1,457	1,193	967	964	2,099	1,028	1,102	4,295	World Bank (https://data.worldbank.org/indicator/NY.GDP.PCAP.CD?end=2022&most_recent_value_desc=true&start=1960&view=chart)
5.2	4.6	8.2	4.6	4.8	5.3	-1.0	6.6	World Bank (https://data.worldbank.org/indicator/NY.GDP.MKTP.KD.ZG)
0.8	0.9	3.6	2.7	3.6	0.4	2.9	5.9	World Bank (https://data.worldbank.org/indicator/BX.TRF.PWKR.DT.GD.ZS)
6.1	2.8	13.0	4.3	5.5	4.0	18.7	7.0	World Bank (https://data.worldbank.org/indicator/SL.UEM.TOTL.ZS)
11.0	4.4	17.7	7.2	7.7	33.9	138.8	13.9	World Bank (https://data.worldbank.org/indicator/FP.CPI.TOTL.ZG)
11.1	-4.2	-16.2	-10.5	-9.3	-10.1	0.4	-6.8	World Bank (https://data.worldbank.org/indicator/NE.RSB.GNFS.ZS)
99.5	44.9	61.1	49.7	68.4	47.1	185.8	88.5	UN Trade and Development (https://unctad.org/publication/world-of-debt)
Presidential republic	Presidential republic	Presidential republic	Presidential republic	Presidential republic	Federal parliamentary republic	Presidential republic	Presidential republic	CIA factbook (https://www.cia.gov/the-world-factbook/field/government-type/)
1964	1964	1962	1962	1963	[2]	1956	1922	Wikipedia (https://en.wikipedia.org/wiki/Decolonisation_of_Africa) [1] **Union of South Africa until 1961** [2] **1935–1941 Italian occupation**

Data Sheet

Thematic area	Indicator	Unit	Ref. year	World	Germany	South Africa	Botswana	Namibia	Angola
Politics	Former colonial power				-	Great Britain	Great Britain	Germany, South Africa	Portugal
	Political stability & human rights (Fragile States Index)		2022	-	23.6 (167/179)	72.0 (79/179)	56.1 (123/179)	62.9 (109/179)	88.1 (35/179)
	Corruption Perceptions Index (CPI)		2022	-	79 (9/180)	43 (72/180)	60 (35/180)	49 (59/180)	33 (116/180)
Environment	Greenhouse gas emissions	[t CO_2e/ person/year]	2022	-	9.49	8.91	5.05	4.67	1.90
	Environmental Performance Index (EPI)		2022	-	62.4 (13/180)	37.2 (116/180)	54.0 (35/180)	50.9 (44/180)	30.5 (151/180)
	Forest area	[% of land area]	2021	31.2	32.7	14.0	26.7	8.0	53.0
Education	Average years of schooling	[years]	2022	-	14.3	11.6	10.4	7.2	5.8
	Education expenditure	[% of GDP]	2022	3,7	4.5	6.2	8.1 (2020)	9.7	2.3
	Literacy rate	[%]	2022	87	99 (2003)	90 (2021)	88 (2015)	92 (2021)	72
Energy sector	Primary energy consumption	[kWh/year/ person]	2021	-	42,497	23,675	9,385	8,748	2,791
	Electricity consumption	[kWh/year/ person]	2021	3,204	6,138	3,216	1,314	1,345	435
	Share of electricity consumption in energy consumption[4]	[%]	2021	-	14	14	14	15	16
	Share of renewables in total electricity generation (excluding hydropower)	[%]	2022	14.5	41.2	8.5	0.4	38.5	0.4
	Share of renewables in total electricity generation (including hydropower)	[%]	2022	29.4	44.3	9.8	0.4	96.3	75.0
	– Wind	[%]	2022	7.3	22.0	4.1	0.0	1.5	0.0
	– Solar	[%]	2022	4.6	10.7	4.2	0.4	37.0	0.1
	– Biomass	[%]	2022	2.3	8.5	0.2	0.0	0.0	0.3
	– Other renewables	[%]	2022	0.3	0.00	0.0	0.0	0.0	0.0
	– Hydropower	[%]	2022	14.9	3.1	1.3	0.0	57.8	74.6
	Share of fossil fuels in total electricity generation	[%]	2022	70.6	55.7	90.2	99.6	3.7	25.0
	– Nuclear	[%]	2022	9.2	6.1	4.2	0.0	0.0	0.0

Zambia	Tanzania	Rwanda	Uganda	Kenya	Ethiopia	Sudan	Egypt	Source/*formula*/comment
Great Britain	Germany, Great Britain	Germany, Great Britain	Great Britain	Great Britain	-	Great Britain	Great Britain	Wikipedia (https://en.wikipedia.org/wiki/Decolonisation_of_Africa)
83.6 (42/179)	78.2 (61/179)	83.7 (41/179)	92.1 (25/179)	88.2 (33/179)	99.3 (13/179)	107.1 (7/179)	83.6 (42/179)	https://fragilestatesindex.org/global-data/
33 (116/180)	38 (94/180)	51 (54/180)	26 (142/180)	32 (123/180)	38 (94/180)	22 (162/180)	30 (130/180)	https://www.transparency.de/cpi/cpi-2022/cpi-2022-tabellarische-rangliste
1.53	1.34	0.66	1.12	2.10	1.63	2.42[3]	3.55	Emissions Database for Global Atmospheric Research (https://edgar.jrc.ec.europa.eu/report_2023?vis=ghgpop#emissions_table) **[3] value reported including South Sudan**
38.4 (107/180)	34.2 (134/180)	32.8 (138/180)	35.8 (125/180)	30.8 (148/180)	31.8 (143/180)	27.6 (171/180)	35.5 (127/180)	https://epi.yale.edu/epi-results/2022/component/epi
60.0	51.1	11.2	11.5	6.3	15.1	9.7	0.0	World Bank (https://data.worldbank.org/indicator/AG.LND.FRST.ZS)
7.3	5.6	4.9	6.2	7.7	2.4	3.9	9.8	United Nations Development Programme (https://hdr.undp.org/data-center/specific-country-data#/countries)
3.6	3.2	4.8	2.6	4.6 (2020)	3.7	2.0 (2009)	3.9 (2015)	World Bank (https://databank.worldbank.org/source/world-development-indicators/Series/SE.XPD.TOTL.GD.ZS)
88 (2020)	82	76 (2021)	81	83	52 (2017)	61 (2018)	75	Our World in Data (https://ourworldindata.org/literacy)
3,419	907	471	767	1,953	872	2,317	9,690	Our World in Data (https://ourworldindata.org/grapher/per-capita-energy-use?tab=table&time=2021..2022)
719	113	52	68	172	79	285	1,538	Wikipedia (https://en.wikipedia.org/wiki/List_of_countries_by_electricity_consumption)
21	12	11	9	9	9	12	16	*Electricity consumption / primary energy consumption* **[4] value calculated using datasets from different sources.**
1.1	1.3	2.0	9.7	67.4	4.3	0.8	5.1	Ember (https://ember-climate.org/data/data-tools/data-explorer/)
88.9	32.6	54.1	98.9	92.0	99.9	62.4	11.9	
0.0	0.0	0.0	0.0	17.3	3.8	0.0	2.56	
0.7	0.6	2.0	2.4	3.1	0.3	0.22	2.50	
0.4	0.8	0.0	7.2	2.4	0.2	0.62	0.00	
0.0	0.0	0.0	0.0	44.7	0.0	0.00	0.00	
87.8	31.3	52.0	89.2	24.6	95.7	61.59	6.87	
11.1	67.4	45.9	1.1	8.0	0.1	37.6	88.1	
0.0	0.0	0.0	0.0	0.0	0.0	0.0	0.0	

Data Sheet

Thematic area	Indicator	Unit	Ref. year	World	Germany	South Africa	Botswana	Namibia	Angola
Energy sector	– Coal	[%]	2022	35.7	31.7	84.4	95.8	3.7	0.0
	– Gas	[%]	2022	22.8	14.1	0.0	0.0	0.0	9.9
	– Other fossils	[%]	2022	3.0	3.8	1.5	3.8	0.0	15.1
Infrastructure	Access to electricity	[% of population]	2022	91	100	87	76	56	49
	Internet users	[% of population]	2021	63	91	72	74	53	33
	Roads: total road length[5]	[km]	2021	-	642,401[a]	750,000[a]	32,564[a]	48,540[a]	76,000 (2019)[a]
	Roads: total road length per 100 km²[5,6]	[km/100 km²]	2021	-	180[a]	61[a]	6[a]	6[a]	6 (2019)[a]
	Roads: total road length per 1,000 inhabitants (population as of 2022)[5,6]	[km/1,000 inhabitants]	2021	-	8[a]	13[a]	13[a]	19[a]	2 (2019)[a]
	Roads: share of paved roads[5]	[%]	2021	-	100 (2022)[b]	22[a]	33[a]	17[a]	18 (2020)[b]
	Road safety: traffic deaths per 100,000 inhabitants	[deaths/ 100,000 inhabitants]	2019	-	4	22	26	35	26
	Railways: route length per 100 km²	[km/100 km²]	2006	-	10.80 (2017)	1.72 (2017)	0.15 (2014)	0.29	0.22
	Railways: route length per 1,000 inhabitants	[km/1,000 inhabitants]	2006	-	0.47 (2017)	0.39 (2017)	0.40 (2014)	1.14	0.14
	Railways: share of electrified line	[%]	2006	-	55 (2017)	47 (2017)	0 (2014)	0	0
	Number of international airports[7]		2022	-	19	8	4	2	2
	Airports: freight transport volume	[million tonne-km]	2019[8]	219,524	7,764	565	1	7	68
	Airports: passenger numbers	[1,000 passengers]	2019[8]	4,455,690	109,634	26,211	254	213	1,437
	Ports: container throughput	[1,000 20-foot standard containers]	2021	840,636	14,716	4,415	-	154	700 (2020)
Production and trade: Overall economy	Total exports	[mil. US$/year]	2022	-	1,686,142	121,616	8,415	5,828	51,275
	Total imports	[mil. US$/year]	2022	-	1,588,853	111,880	8,115	7,795	17,803
	Trade balance	[mil. US$/year]	2022	-	97,289	9,736	300	-1967	33,472

Zambia	Tanzania	Rwanda	Uganda	Kenya	Ethiopia	Sudan	Egypt	Source/*formula*/comment
11.1	0.0	5.10	0.0	0.0	0.0	0.0	0.0	Ember (https://ember-climate.org/data/data-tools/data-explorer/)
0,0	67.2	24.5	0.0	0.0	0.0	0.0	79.3	
0,0	0.2	16.3	1.1	8.0	0.1	37.6	8.8	
48	46	51	47	76	55	63	100	World Bank (https://data.worldbank.org/indicator/EG.ELC.ACCS.ZS)
21	32	30	10	29	17	28 (2020)	72	World Bank (https://data.worldbank.org/indicator/IT.NET.USER.ZS)
67,305[a]	180,791[a]	6,674[a]	146,000 (2019)[a]	161,833[a]	155,830[a]	30,000 (2019)[b]	150,052 (2020)[a]	[a] International Road Federation (https://datawarehouse.worldroadstatistics.org/dashboard/linegraph) [b] CIA Factbook (https://www.cia.gov/the-world-factbook/field/roadways/) [5] **Road network figures vary (sometimes substantially) between different sources.**
9[a]	19[a]	25[a]	60 (2019)[a]	28[a]	14[a]	2 (2019)[b]	15 (2020)[a]	*Total road length / land area * 100* [6] **Value calculated using datasets from different sources.**
3[a]	3[a]	0.5[a]	3 (2019)[a]	3[a]	1[a]	1 (2019)[b]	1 (2020)[a]	*Total road length in km / population * 1.000*
16[a]	8[a]	26[a]	-	13[a]	12[a]	27 (2019)[b]	68[a]	[a] International Road Federation (https://datawarehouse.worldroadstatistics.org/dashboard/linegraph) [b] CIA Factbook (https://www.cia.gov/the-world-factbook/field/roadways/)
21	31	29	29	28	28	27	10	World Bank (https://data.worldbank.org/indicator/SH.STA.TRAF.P5)
0.16	0.29	0.00	0.11 (2002)	0.44 (2013)	0.06 (2016)	0.29	0.65 (2017)	Wikipedia(https://en.wikipedia.org/wiki/List_of_countries_by_rail_transport_network_size)
0.09	0.06	0.00	0.01 (2002)	0.06 (2013)	0.01 (2016)	0.18	0.07 (2017)	
0	8	0	0 (2002)	0 (2013)	64 (2016)	0	1 (2017)	
4	6	1	3	4	2	2	16	Wikipedia(https://en.wikipedia.org/wiki/List_of_international_airports_by_country) [7] **Figures vary between different sources.**
79	1	0	0	321	2,450	0	483	World Bank (https://data.worldbank.org/indicator/IS.AIR.GOOD.MT.K1?end=2019&start=1970) [8] **Reference year 2019 chosen because passenger numbers dropped sharply due to the Covid-19 pandemic.**
18	1,498	1,562	20	6,418	12,631	530	13,207	World Bank (https://data.worldbank.org/indicator/IS.AIR.PSGR?end=2019&start=1970)
-	710 (2020)	-	-	1,436	-	470 (2019)	7,322	World Bank (https://data.worldbank.org/indicator/IS.SHP.GOOD.TU?end=2021&skipRedirection=true&start=2000&view=chart)
11,643	6,825	2,019	4,149 (2020)	7,371	3,085	3,619 (2018)	52,116	UN Comtrade Database (https://comtradeplus.un.org/TradeFlow)
9,004	15,654	5,125	8,251 (2020)	21,101	16,538	10,484 (2018)	96,189	
2,639	-8,830	-3,105	-4,102 (2020)	-13,730	-13,453	-6,865 (2018)	-44,073	

Data Sheet

Thematic area	Indicator	Unit	Ref. year	World	Germany	South Africa	Botswana	Namibia	Angola
Overall economy	Main export partner (1st)		2022		USA	China	United Arab Emirates	Botswana	China
		[export share in %]	2022		10	10	27	17	43
		[export share in mil. US$]	2022		165,538	11,695	2,282	998	21,905
	Main export partner (2nd)		2022		France	USA	Belgium	South Africa	India
		[export share in %]	2022		7	9	19	15	10
		[export share in mil. US$]	2022		124,071	10,728	1,596	855	5,141
	Main export partner (3rd)		2022		Netherlands	Germany	India	China	France
		[export share in %]	2022		7	7	15	12	7
		[export share in mil. US$]	2022		116,621	8,835	1,280	720	3,652
	Main import partner (1st)		2022		China	China	South Africa	South Africa	China
		[import share in %]	2022		13	20	63	36	16
		[import share in mil. US$]	2022		206,642	22,462	5,085	2,794	2,856
	Main import partner (2nd)		2022		Netherlands	India	Namibia	China	Portugal
		[import share in %]	2022		7	7	7	7	11
		[import share in mil. US$]	2022		113,437	8,334	578	547	1,913
	Main import partner (3rd)		2022		USA	Germany	Belgium	India	Republic of Korea
		[import share in %]	2022		6	7	6	5	9
		[import share in mil. US$]	2022		99,761	8,205	470	358	1,642
Cement & concrete	Number of active cement plants		2023	-	52	20	1	2	6
	Cement production capacity	[mil. t/year]	2023	-	77.2	30.0	0.4	2.5	9.5
	Exports cement & concrete	[mil. US$/year]	2022	-	1,109	19.3	5.7	0.1	0.0
	Imports cement & concrete	[mil. US$/year]	2022	-	511	31.1	2.8	1.7	2.5
	Trade balance cement & concrete	[mil. US$/year]	2022	-	598	-11.7	2.9	-1.6	-2.4

Zambia	Tanzania	Rwanda	Uganda	Kenya	Ethiopia	Sudan	Egypt	Source/*formula*/comment
Switzerland	India	Democratic Republic of the Congo	United Arab Emirates (2020)	Uganda	USA	United Arab Emirates (2018)	Turkey	UN Comtrade Database (https://comtradeplus.un.org/TradeFlow)
39	17	38	44 (2020)	11	11	27	8	
4,487	1,179	768	1,845 (2020)	820	335	994 (2018)	3,960	
China	South Africa	United Arab Emirates	Kenya (2020)	USA	Saudi Arabia	China (2018)	Spain	
21	14	29	11	9	9	20	7	
2,407	930	590	466 (2020)	679	268	736 (2018)	3,850	
Democratic Republic of the Congo	United Arab Emirates	China	South Sudan (2020)	Netherlands	Somalia	Saudi Arabia (2018)	Italy	
14	11	5	9	8	9	16	6	
1,589	763	97	357 (2020)	591	265	588 (2018)	3,382	
South Africa	China	China	China (2020)	China	China	China (2018)	China	
30	25	21	16	18	30	17	15	
2,705	3,945	1,078	1,352 (2020)	3,840	4,960	1,823 (2018)	14,764	
China	United Arab Emirates	Tanzania	India (2020)	United Arab Emirates	India	Russia (2018)	Saudi Arabia	
15	16	11	12	16	14	15	8	
1,387	2,492	568	959 (2020)	3,456	2,366	1,582 (2018)	7,886	
United Arab Emirates	India	Kenya	Kenya (2020)	India	USA	Saudi Arabia (2018)	USA	
8	13	9	9	10	7	10	7	
684	1,961	479	773 (2020)	2,121	1,103	1,071 (2018)	6,953	
6	7	1	4	10	7	6	24	https://datadryad.org/stash/dataset/doi:10.5061/dryad.6tlgljx4f
5.5	11.5	0.6	3.3	11.0	14.6	8.6	76.1	https://datadryad.org/stash/dataset/doi:10.5061/dryad.6tlgljx4f HS-Codes (https://de.abrams.wiki/hscode/6810)
0.5	0.0	0.0	2.7 (2020)	0.1	0.0	0.0 (2018)	0.7	UN Comtrade Database (https://comtradeplus.un.org/TradeFlow (Product code HS 6810)
1.4	1.5	1.0	0.1 (2020)	1.4	2.0	0.1 (2018)	4.2	
-0.9	-1.5	-1.0	2.6	-1.3	-2.0	-0.1	-3.5	

Data Sheet

Thematic area	Indicator	Unit	Ref. year	World	Germany	South Africa	Botswana	Namibia	Angola
Cement & concrete	Main export partner (1st)		2022		Netherlands	Lesotho	Zimbabwe	No significant exports	No significant exports
		[export share in %]	2022		21	21	68		
		[export share in mil. US$]	2022		229.8	4.1	3.9		
	Main export partner (2nd)		2022		France	Botswana	South Africa		
		[export share in %]	2022		11	14	30		
		[export share in mil. US$]	2022		126.8	2.6	1.7		
	Main export partner (3rd)		2022		Austria	Zimbabwe	-		
		[export share in %]	2022		11	13	-		
		[export share in mil. US$]	2022		119.5	2.5	-		
	Main import partner (1st)		2022		Czech Republic	China	South Africa	USA	Great Britain
		[import share in %]	2022		23	65	94	46	32
		[import share in mil. US$]	2022		119.7	20.1	2.6	0.8	0.8
	Main import partner (2nd)		2022		Poland	Israel	China	South Africa	China
		[import share in %]	2022		21	10	5	34	29
		[import share in mil. US$]	2022		105.6	3.1	0.2	0.6	0.7
	Main import partner (3rd)		2022		Netherlands	Botswana	-	China	United Arab Emirates
		[import share in %]	2022		11	6	-	13	13
		[import share in mil. US$]	2022		58.6	1.8	-	0.2	0.3
Iron & steel	Steel production (crude steel)	[1,000 t/year]	2022	-	36,860	4,403	-	-	270
	Steel production capacity (crude steel)	[1,000 t/year]	2022	-	58,100	8,055	86	1,000	500
	Steel exports	[mil. US$/year]	2022	-	37,674	6,579	14	38	32
	Steel imports	[mil. US$/year]	2022	-	39,576	1,814	94	124	261
	Steel trade balance	[mil. US$/year]	2022	-	-1,902	4,765	-80	-86	-229

Zambia	Tanzania	Rwanda	Uganda	Kenya	Ethiopia	Sudan	Egypt	Source/*formula*/comment
Democratic Republic of the Congo	No significant exports	No significant exports	Kenya (2020)	No significant exports	No significant exports	No significant exports	Libya	UN Comtrade Database (https://comtradeplus.un.org/TradeFlow (Product code HS 6810). **Only bilateral trade flows of at least US$100.000 are included.**
69			93				42	
0.4			2.6 (2020)				0.3	
Democratic Republic of the Congo			South Sudan (2020)				Sudan	
29			7				27	
0.1			0.2 (2020)				0.2	
-			-					
-			-					
-			-					
China	Turkey	China	No significant imports	India	China	No significant imports	Greece	UN Comtrade Database (https://comtradeplus.un.org/TradeFlow (Product code HS 6810). **Only bilateral trade flows of at least US$100.000 are included.**
32	47	86		26	60		34	
0.4	0.7	0.8		0.4	1.2		1.4	
South Africa	China	-		Israel	Israel		Saudi Arabia	
31	24	-		23	18		31	
0.4	0.4	-		0.3	0.4		1.3	
Zimbabwe	India	-		China	Italy		China	
20	12	-		18	8		19	
0.3	0.2	-		0.3	0.2		0.8	
-	140	-	-	450	0	0	9,819	World Steel (https://worldsteel.org/steel-topics/statistics/annual-production-steel-data/?ind=P1_crude_steel_total_pub/CHN/IND)
100	-	-	72	645	760	60	15,200	OECD (https://stats.oecd.org/Index.aspx?datasetcode=STI_STEEL_MAKINGCAPACITY) **Values for steel production and steel production capacity from World Steel and the OECD partly contradict each other.**
328	77	34	68 (2020)	202	1	0 (2018)	1,409	UN Comtrade Database (https://comtradeplus.un.org/TradeFlow (Product code HS 72))
294	851	200	360 (2020)	1.237	569	189 (2018)	5,133	
34	-774	-165	-292	-1,036	-568	-189	-3,724	

Data Sheet

Thematic area	Indicator	Unit	Ref. year	World	Germany	South Africa	Botswana	Namibia	Angola
Iron & steel	Main export partner (1st)		2022		France	China	South Africa	United Arab Emirates	Democratic Republic of the Congo
		[export share in %]	2022		10	20	88	45	71
		[export share in mil. US$]	2022		3,803	1,318	13	17	23
	Main export partner (2nd)		2022		Poland	Mozambique	-	Luxembourg	Ghana
		[export share in %]	2022		10	17	-	16	17
		[export share in mil. US$]	2022		3,741	1,104	-	6	5
	Main export partner (3rd)		2022		Italy	USA	-	Zambia	Senegal
		[export share in %]	2022		10	11	-	15	11
		[export share in mil. US$]	2022		3,719	719	-	6	4
	Main import partner (1st)		2022		Italy	China	South Africa	South Africa	China
		[import share in %]	2022		13	41	96	51	64
		[import share in mil. US$]	2022		5,009	751	90	63	168
	Main import partner (2nd)		2022		Belgium	Germany	China	China	Portugal
		[import share in %]	2022		13	9	3	45	12
		[import share in mil. US$]	2022		4,958	172	3	55	32
	Main import partner (3rd)		2022		France	Japan	-	Zambia	Italy
		[import share in %]	2022		12	6	-	3	8
		[import share in mil. US$]	2022		4,755	115	-	4	20
Raw timber	Forest area	[% of land area]	2021	31.2	32.7	14.0	26.7	8.0	53.0
	Roundwood exports	[mil. US$/year]	2022	-	1,142	85.1	0.0	2.8	0.0
	Roundwood imports	[mil. US$/year]	2022	-	578	5.2	12.5	9.0	0.0
	Roundwood trade balance	[mil. US$/year]	2022	-	565	79.9	-12.5	-6.2	0.0

Zambia	Tanzania	Rwanda	Uganda	Kenya	Ethiopia	Sudan	Egypt	Source/*formula*/comment
Italy	Uganda	Democratic Republic of the Congo	South Sudan (2020)	Uganda	No significant exports	No significant exports	Spain	UN Comtrade Database (https://comtradeplus.un.org/TradeFlow (Product code HS 72)) **Only bilateral trade flows of at least US$1.000.000 are included.**
24	29	96	30	42			19	
77	23	33	21 (2020)	84			268	
Democratic Republic of the Congo	Democratic Republic of the Congo	Zambia	Tanzania (2020)	Tanzania			Saudi Arabia	
20	27	4	25	24			14	
67	21	1	17 (2020)	49			204	
South Africa	Burundi	-	Democratic Republic of the Congo (2020)	Democratic Republic of the Congo			Italy	
11	18	-	21	14			13	
35	14	-	15 (2020)	28			189	
South Africa	China	China	Japan (2020)	China	China	China (2018)	Great Britain	UN Comtrade Database (https://comtradeplus.un.org/TradeFlow (Product code HS 72)) **Only bilateral trade flows of at least US$1.000.000 are included.**
44	42	64	25	29	45	52	13	
129	355	128	90 (2020)	357	256	99 (2018)	654	
China	Japan	Kenya	China (2020)	Japan	Turkey	Egypt (2018)	China	
35	17	14	19	21	19	18	12	
102	143	29	69 (2020)	264	109	35 (2018)	625	
Chile	South Africa	Malawi	Kenya (2020)	India	Ukraine	Iran (2018)	Turkey	
11	15	6	19	17	12	12	11	
32	127	11	68 (2020)	215	67	22 (2018)	540	
60.0	51.1	11.2	11.5	6.3	15.1	9.7	0.0	World Bank (https://data.worldbank.org/indicator/AG.LND.FRST.ZS)
11.2	5.1	0.0	0.1 (2020)	0.5	0.1	0.0 (2018)	0.0	UN Comtrade Database (https://comtradeplus.un.org/TradeFlow (Product code HS 4403)
3.5	2.3	0.7	0.0 (2020)	0.5	0.3	0.2 (2018)	49.3	
7.7	2.8	-0.7	0.1	0.0	-0.1	-0.2	-49.3	

Data Sheet

Thematic area	Indicator	Unit	Ref. year	World	Germany	South Africa	Botswana	Namibia	Angola
Raw timber	Main export partner (1st)		2022		China	China	No significant exports	South Africa	No significant exports
		[export share in %]	2022		42	29		70	
		[export share in mil. US$]	2022		476.9	24.5		2.0	
	Main export partner (2nd)		2022		Austria	India		Vietnam	
		[export share in %]	2022		22	10		18	
		[export share in mil. US$]	2022		255.9	8.3		0.5	
	Main export partner (3rd)		2022		Belgium	Botswana		China	
		[export share in %]	2022		7	9		6	
		[export share in mil. US$]	2022		82.8	7.6		0.2	
	Main import partner (1st)		2022		Czech Republic	Eswatini	South Africa	South Africa	No significant imports
		[import share in %]	2022		32	97	88	91	
		[import share in mil. US$]	2022		184.0	5.1	11.0	8.2	
	Main import partner (2nd)		2022		Poland	-	Zimbabwe	Zambia	
		[import share in %]	2022		17	-	7	7	
		[import share in mil. US$]	2022		99.7	-	0.9	0.6	
	Main import partner (3rd)		2022		Norway	-	Eswatini	Angola	
		[import share in %]	2022		13	-	4	1	
		[import share in mil. US$]	2022		75.4	-	0.5	0.1	

Zambia	Tanzania	Rwanda	Uganda	Kenya	Ethiopia	Sudan	Egypt	Source/*formula*/comment
Vietnam	China	No significant exports	No significant exports	Rwanda	Somalia	No significant exports	No significant exports	UN Comtrade Database (https://comtradeplus.un.org/TradeFlow (Product code HS 4403) **Only bilateral trade flows of at least US$ 100.000 are included.**
81	79			80	99			
9.1	4.1			0.4	0.1			
China	Rwanda			-	-			
19	10			-	-			
2.1	0.5			-	-			
-	Zambia			-	-			
-	3			-	-			
-	0.2			-	-			
South Africa	South Africa	Uganda	No significant imports	Uganda	Italy	Egypt (2018)	Finland	UN Comtrade Database (https://comtradeplus.un.org/TradeFlow (Product code HS 4403) **Only bilateral trade flows of at least US$ 100.000 are included.**
51	77	100		97	47	89	80	
1.8	1.8	0.7		0.5	0.1	0.2 (2018)	39.2	
Zimbabwe	Uganda	-		-	China	-	Belgium	
47	12	-		-	46	-	9	
1.6	0.3	-		-	0.1	-	4.5	
-	Japan	-		-	-	-	Estonia	
-	11	-		-	-	-	4	
-	0.3	-		-	-	-	1.9	

Literature

Introduction

1. Barbara Käser: “Partnerschule Afrika.” Karl-Mauch-Schule, 2022, www.karl-mauch-schule.de/partnerschule-afrika/.
2. Friedrich Wilhelm Mader: “Am Kilimandjaro.” Union Deutsche Verlagsgesellschaft, Stuttgart, 1927.
3. Friedrich Wilhelm Mader: “Die Flucht aus dem Sudan.” Union Deutsche Verlagsgesellschaft, Stuttgart, 1925.
4. Falk Jäger: “3 Stadia 2010: Architecture for an African Dream.” Jovis, Berlin, 2010.
5. “General History of Africa.” UNESCO, 2024, www.unesco.org/en/general-history-africa.
6. John Reader: “Africa: A Biography of the Continent.” Penguin, London, 1998.
7. Kevin Shillington: “History of Africa.” Bloomsbury, London, 2018.
8. Alan Gladney Russell: “Colour, Race and Empire.” Victor Gollancz, London, 1944.
9. Walter Rodney: “How Europe Underdeveloped Africa.” Bogle-L’Ouverture Publications, London, 1972.
10. Michael Falser (Hg.): “Deutsch-koloniale Baukulturen.” Dietmar Klinger Verlag, Passau, 2023.
11. Geoff Crowther: “Africa on the Cheap.” Lonely Planet, Victoria, 1982.
12. Philipp Meuser und Adil Dalbai (Hg.): “Architectural Guide Sub-Saharan Africa.” DOM publishers, Berlin, 2021.
13. Manuel Herz et al.: “African Modernism.” Park Books, Zürich, 2022, p. 9.
14. “Biennale Architettura 2023: The Laboratory of the Future.” La Biennale di Venezia, 2023, www.labiennale.org/en/news/biennale-architettura-2023-laboratory-future.
15. David P. Billington: “The Tower and the Bridge: The New Art of Structural Engineering.” Princeton University Press, 1986.
16. Wangari Maathai: “The Challenge for Africa.” Penguin, London, 2009, p. 38.
17. Martin Schoeller und Daniel Schönwitz: “Afrika First! Die Agenda für unsere gemeinsame Zukunft.” Berg & Feierabend, Berlin, 2020.
18. Paul Collier: “Exodus: How Migration is Changing Our World.” Oxford University Press, 2015.
19. “De Beers.” Wikipedia, 2024, de.wikipedia.org/wiki/De_Beers.
20. “Kongokonferenz.” Wikipedia, 2024, de.wikipedia.org/wiki/Kongokonferenz.
21. “Sambia.” Wikipedia, 2024, de.wikipedia.org/wiki/Sambia.
22. Ulrich van der Heyden: “Carl Mauchs Aufenthalt im südlichen Afrika und seine Suche nach dem sagenumwobenen Land Ophir.” In: Hannelore van Ryneveld und Jania Wozniak (Hg.): Einzelgang und Rückkehr im Wandel der Zeit: Festschrift für Gunther Pakendorf, SunPress, Stellenbosch, 2010, p. 35–64.
23. Mike Main und Thomas N. Huffman: “Palaces of Stone: Uncovering Ancient Southern African Kingdoms.” Struik Travel & Heritage, Cape Town, 2021.
24. Andreas Eckert: “Geschichte der Sklaverei. Von der Antike bis ins 21. Jahrhundert.” C. H. Beck, München, 2021.
25. “Tanganjikabahn.” Wikipedia, 2024, de.wikipedia.org/wiki/ Tanganjikabahn.
26. “Liemba (Schiff).” Wikipedia, 2024, de.wikipedia.org/wiki/ Liemba_(Schiff).
27. Edward Denison et al.: “Asmara – Africa’s Secret Modernist City.” Merrell, London, 2006.
28. “Bahnstrecke Massaua–Biscia.” Wikipedia, 2024, de.wikipedia.org/wiki/Bahnstrecke_Massaua–Biscia.
29. William Easterly und Ross Levine: “Africa’s Growth Tragedy: A Retrospective, 1960–89.” World Bank, 1995, p. 28.
30. Marc von Lüpke-Schwarz: “Plattenbauten für Sansibar.” DW, 2013, www.dw.com/de/als-die-ddr-der-insel-sansibar-plattenbauten-schenkte/a-17307568.
31. Omer Siddig Osman et al.: “Sudan: Introduction.” Architectural Guide Sub-Saharan Africa, Volume 4, DOM publishers, Berlin, 2021, p. 57.
32. Philipp Meuser und Adil Dalbai: “Theorising Architecture in Sub-Saharan Africa. Perspectives, Questions, and Concepts.” DOM publishers, Berlin, 2021, p. 60.
33. “50 Influential African Women Architects.” Africans Column, 2024, www.africanscolumn.com/50-influential-african-women-architects/.
34. Peter Scholl-Latour: “Afrikanische Totenklage: Der Ausverkauf des Schwarzen Kontinents.” C. Bertelsmann, München, 2001.
35. Peter Scholl-Latour: “Mord am großen Fluß: Ein Vierteljahrhundert afrikanische Unabhängigkeit.” DVA, Stuttgart, 1986.

Diary
South Africa

1. “Südafrika.” Wikipedia, 2024, de.wikipedia.org/wiki/Südafrika#Demografie.
2. “2022 KwaZulu-Natal floods.” Wikipedia, 2024, en.wikipedia.org/wiki/2022_KwaZulu-Natal_floods.

3. Irma Venter: "Msikaba Bridge Heads for Completion by End-2024." Creamer Media's Engineering News, 2023, www.engineeringnews.co.za/article/msikaba-bridge-heads-for-completion-by-end-2024-2023-02-13.

Botswana

1. "Botswana." Wikipedia, 2024, de.wikipedia.org/wiki/Botswana.
2. Martin Grassl und Thomas Benz: "Okavango River Bridge in Botsuana – steile Zähne am Delta." Bautechnik 99.4, 2022, p. 272–279, doi.org/10.1002/bate.202200004.

Namibia

1. Klaus Dierks: "The History of Roads in the South African Era." Klaus Dierks, 2005, www.klausdierks.com/Namibian_Roads/4.htm.
2. Klaus Stratmann: "Namibia startet milliardenschweres Wasserstoff-Projekt mit deutscher Beteiligung", Handelsblatt, 2023, www.handelsblatt.com/politik/international/klimaneutralitaet-namibia-startet-milliardenschweres-wasserstoff-projekt-mit-deutscher-beteiligung/29167756.html.
3. Ulrich Ender: "Heinrich Brockmann." Namibiana Buchdepot, 2014, www.namibiana.de/namibia-information/who-is-who/autoren/infos-zur-person/heinrich-brockmann.html.
4. Nina Maritz und Livingstone Mukasa: "Building a Country and a Profession." Architectural Guide Sub-Saharan Africa, Vol. 7, DOM publishers, Berlin, 2021, p. 268.
5. Yuca Meubrink: "Anders Bauen – Für Ressourcenschonung und Klimaschutz." Berlin-Brandenburgische Akademie der Wissenschaften, 2024, www.bbaw.de/forschung/anders-bauen.
6. Martin Kalb: "Vom Meer ins Land hinein! Die Landungsbrücken von Swakopmund/Deutsch-Südwestafrika (1905–1911)." Deutsch-koloniale Baukulturen, Dietmar Klinger Verlag, Passau, 2023, p. 50.
7. Walter Peters: "Die Otavibahn in Deutsch-Südwestafrika." Deutsch-koloniale Baukulturen, Dietmar Klinger Verlag, Passau, 2023, p. 178.
8. Klaus Dierks: "Die Geschichte der Zambezi-Brücke von Namibia nach Zambia: 1982–2004." Klaus Dierks, 2004, www.klausdierks.com/Strassen/Zambezi_Bruecke.htm.
9. Michael Hofmann: "Militärische Wehrbauten als Postkartenmotive? Zur pittoresken Architektur der Schutztruppen-Festungen in Deutsch-Südwestafrika." Deutsch-koloniale Baukulturen, Dietmar Klinger Verlag, Passau, 2023, p. 258.
10. "BUW-NUST-Kooperation." Bergische Universität Wuppertal, 2019, www.zim.uni-wuppertal.de/de/wir-ueber-uns/projekte/buw-nust-kooperation/.

Angola

1. Wangari Maathai: "The Challenge for Africa." Penguin, London, 2009, S. 85.
2. Onejoon Che: "North Korea's Monumental Gifts to Africa." Architectural Guide Sub-Saharan Africa, Volume 7, DOM publishers, Berlin, 2021, p. 289.
3. "A Histórica Solidariedade da Coreia do Norte com a África." Youtube, uploaded by Juche TV, 2018, www.youtube.com/watch?v=eCYYGkWwHdg&t=230s.
4. "Angola." Wikipedia, 2024, de.wikipedia.org/wiki/Angola.
5. "Expeditions." Angola Uncharted Safari Expeditions, 2020, www.angola-uncharted-safari.com/.
6. Brian Carisma und Veronica Boero: "Government Expenditures in Agriculture 2001–2022." FAO, Rom, 2023, doi.org/10.4060/cc9148en.
7. "Agriculture in Angola." Wikipedia, 2024, en.wikipedia.org/wiki/Agriculture_in_Angola.
8. "Ernährungssicherung in Sambia." GIZ, Bonn, 2022, www.giz.de/de/downloads/giz2022-de-ernaehrungssicherung-in-sambia.pdf.
9. "Zambia's Smallholder Farmers and the Complexities of Government Support." African Farming, 4. Januar 2024, https://web.archive.org/web/20240918054840/www.africanfarming.com/top-stories/zambias-smallholder-farmers-complexities-government-support/.

Zambia

1. Fabian Urech: "Eine Brücke, die sich um das widerspenstige Nachbarland krümmt: Simbabwe bezahlt teuer für Robert Mugabes Eigensinn." Neue Zürcher Zeitung, 3. März 2023, www.nzz.ch/international/wie-eine-grenzbruecke-in-afrika-zum-diplomatischen-zankapfel-wurde-ld.1727991.
2. "Victoria Falls Bridge." Wikipedia, 2024, de.wikipedia.org/wiki/Victoria_Falls_Bridge.
3. Philipp Meuser und Adil Dalbai: "Zambia: Buildings and Projects." Architectural Guide Sub-Saharan Africa, Volume 7, DOM publishers, Berlin, 2021, p. 154.
4. Francis Mwape Ndilia: "Housing in Zambia: Challenges and Opportunities." Architectural Guide Sub-Saharan Africa, Volume 7, DOM publishers, Berlin, 2021, p. 158.
5. Haitham Al Abdulrazzak: "Bauen in Afrika – Tansania." Bachelorarbeit, TU Berlin: Fachgebiet Entwerfen und Konstruieren – Massivbau, 2023.
6. "Housing." Tanzania Invest, 2020, www.tanzaniainvest.com/housing.
7. Conrad Simuchile: "Chambeshi Railway Bridge Now Open to Traffic." Tanzania-Zambia Railway Authority, 2022, www.tazarasite.com/chambeshi-railway-bridge-now-open-traffic.
8. Paul von Lettow-Vorbeck: "Heia Safari! Deutschlands Kampf in Ostafrika." Koehler, Biberach, 1952.
9. "Shiwa Ngandu." Wikipedia, 2024, en.wikipedia.org/wiki/Shiwa_Ngandu.

Tanzania

1. Miguel d'Escoto Brockmann: "Address at the United Nations General Assembly." 2009, New York. UN General Assembly, https://www.un.org/esa/socdev/social/intldays/IntlJustice/launch10Feb09/PGA.pdf.
2. "Closed Compact Report: Tanzania Compact." Millennium Challenge Corporation, 2015, www.mcc.gov/where-we-work/program/tanzania-compact/.
3. "Trunk and Regional Roads Network." Tanroads, 2018, www.tanroads.go.tz/uploads/files/Trunk_and_Regional_Road_Map.pdf.
4. "Tanzania: Halt Brutal Security Operation in Loliondo." Amnesty International, 15. Juni 2022, www.amnesty.org/en/latest/news/2022/06/tanzania-halt-brutal-security-operation-in-loliondo/.

Rwanda

1. "Ruanda." Wikipedia, 2024, de.wikipedia.org/wiki/Ruanda.
2. "Coltan." Wikipedia, 2024, en.wikipedia.org/wiki/Coltan.
3. "Improving East Africa's Inland Logistics." We Are Tomodachi, 2016, www.japan.go.jp/tomodachi/_userdata/pdf/2016/japan_in_africa_edition_2016/08_09.pdf.
4. "Sprachen Ruandas." Wikipedia, 2024, de.wikipedia.org/wiki/Sprachen_Ruandas.
5. "SAICE-Africa." SAICE-Africa, 2019, saice-africa.org.
6. "Cutting slates from Volcanic rocks." YouTube, Rising Engineer, 2020, https://www.youtube.com/watch?v=adRHG6-16rw.

Uganda
1. "City Population 2100." OntarioTech University, 2024, sites.ontariotechu.ca/sustainabilitytoday/urban-and-energy-systems/Worlds-largest-cities/population-projections/city-population-2100.php.
2. "Makerere-Universität." Wikipedia, 2020, de.wikipedia.org/wiki/Makerere-Universität.
3. "Uganda Makerere University Fire: ‚Ivory Tower' Gutted." BBC, 2020, www.bbc.com/news/world-africa-54225111.
4. "A Map of Jinja's Historical Buildings and Sites." The Cross-Cultural Foundation of Uganda, 2019, www.crossculturalfoundation.or.ug/wp-content/uploads/2021/03/Jinja-Historical-Buildings-and-Sites-Map-May-2019.pdf.
5. Lonely Planet: "East Africa." Lonely Planet, London, 2018, p. 476.
6. Joseph Kamau und Andrew Cameron: "Lust to Kill, the Rise and Fall of Idi Amin." Corgi Book, London, 1979.

Kenya
1. Simon Wanyonyi Otando: "An Engineering Geological Characterisation of Tropical Clays. Case Study: Clay Soils of Nairobi, Kenya." Dissertation, Technische Universität Clausthal, 2004, p. 26, dokumente.ub.tu-clausthal.de/servlets/MCRFileNodeServlet/rimport_derivate_00000040/db107471.pdf.
2. "A Study on the Engineering Behaviour of Nairobi Subsoil." Langeo, https://web.archive.org/web/20250624172758/http://www.langeoinstrument.com/case/117.html.
3. E. P. Saggerson: "Geology of the Nairobi Area." Ministry of Environment and Natural Resources: Mines and Geological Department, Nairobi, 1991, S. 77, https://s3-eu-west-1.amazonaws.com/samsamwater1/maps/kenya/geology/Geology+of+the+Nairobi+area.pdf.
4. Xuxue Jiao et. al.: "Influence of Kenyan Phonolite Manufactured Sand on Concrete Performance." 3rd Global Conference on Ecological Environment and Civil Engineering, Guangzhou, China, 2021, doi.org/10.1051/e3sconf/202129302045.
5. "Lawrence Hall, London." Wikipedia, 2023, en.wikipedia.org/wiki/Lawrence_Hall,_London.
6. Albert Andeso: "Work Set to Begin on Sh85bn Sea Bridge in Mombasa." Construction Kenya, 15. März 2022, www.constructionkenya.com/9077/mombasa-gate-bridge-masterplan/.
7. Pascal Heidinger: "Bauen in Afrika – Kenia." Bachelorarbeit, TU Berlin: Fachgebiet Entwerfen und Konstruieren – Massivbau, 2023.
8. "Athi River Super Bridge." Wikipedia, 2022, en.wikipedia.org/wiki/Athi_River_Super_Bridge.

Ethiopia
1. "Wirtschaft Äthiopiens." Wikipedia, 2023, de.wikipedia.org/wiki/Wirtschaft_Äthiopiens.
2. Jürgen Zurheide: "Der vergessene Krieg in Äthiopien." Deutschlandfunk, 15. April 2023, www.deutschlandfunk.de/interview-navid-kermani-zu-besuch-in-tigray-und-amhara-erschreckende-bilanz-dlf-a79e647f-100.html.
3. "What's behind the Crisis in Ethiopia's Amhara Region? A Simple Guide." Aljazeera, 10. August 2023, www.aljazeera.com/news/2023/8/10/whats-behind-the-crisis-in-ethiopias-amhara-region-a-simple-guide.
4. Teshome Borago: "Ethiopia: What Is the African #NoMore Movement about?" Borkena Ethiopian News, 21. November 2021, www.borkena.com/2021/11/21/nomore-movement-about-ethiopia-what-is-the-african/.
5. "Foreign Aid: Holds Itself Back." Ethiopian Business Review, Jan. 2022, www.ethiopianbusinessreview.net/foreign-aid-holds-itself-back/.
6. Ilona Eveleens: "Wasserversorgung in Addis Abeba: Eine Stadt vergiftet sich selbst." taz, 1. September 2021, www.taz.de/Wasserversorgung-in-Addis-Abeba/!5796599/.
7. Siegbert Uhlig et al.: "Äthiopien. Geschichte, Kultur, Herausforderungen." Harrassowitz, Wiesbaden, 2018, p. 82, 340 and 364.
8. Lars Krückeberg, Wolfram Putz und Thomas Willemeit: "Empowering Villages with Self-build Kits." Architectural Guide Sub-Saharan Africa, Volume 4, DOM publishers, Berlin, 2021, p. 342.
9. Ayala Levin: "Exporting Zionism: Architectural Modernism in Israeli-African Technical Cooperation, 1958–1973." Dissertation, Columbia University, 2015, p. 352 and 357–359.
10. Georg Gerster: "Kirchen im Fels. Entdeckungen in Äthiopien." Kohlhammer, Stuttgart, 1968.
11. Fasil Giorghis: "City Portrait: Gonder." Architectural Guide Sub-Saharan Africa, Vol. 4, DOM publishers, Berlin, 2021, p. 257.
12. Meheret Selassie Mokonnen: "Abandoned Historical Bridges." The Reporter, 7. Okt. 2017, www.thereporterethiopia.com/1166/.

Sudan
1. "Charles George Gordon." Wikipedia, 2024, de.wikipedia.org/wiki/Charles_George_Gordon.
2. Ibrahim Z. Bahreldin: "Gordon Memorial College." Architectural Guide Sub-Saharan Africa, Volume 4, DOM publishers, Berlin, 2021, p. 82.
3. Omer S. Osman und Amira O.S. Osman: "The modern heritage of Khartoum." docomomo 32, 2005, p. 68, www.docomomojournal.com/index.php/journal/issue/view/54/27.
4. "Alte Omdurman-Brücke." Wikipedia, 2022, de.wikipedia.org/wiki/Alte_Omdurman-Brücke.
5. Wolfram Haumer et al.: "Die Instandsetzung der Old White Nile Bridge in Khartoum/Sudan." Stahlbau 79.12, 2010; S. 883–891, doi.org/10.1002/stab.201001382.
6. "Tuti Suspension Bridge, Khartoum." DLT Engineering, 2019, https://web.archive.org/web/20241207022243/www.dormanlongtechnology.com/en/projects/Tuti.htm.
7. "Sudanese Army Accuses RSF of Destroying Strategic Bridge in the Capital." Sudan Tribune, 2023, https://web.archive.org/web/20250724064320/https://sudantribune.com/article279244/.
8. "Old Dongola." Wikipedia, 2024, en.wikipedia.org/wiki/Old_Dongola.
9. Ibrahim Z. Bahreldin et al.: "The Work of Abdallah Sabbar." Architectural Guide Sub-Saharan Africa, Volume 4, DOM publishers, Berlin, 2021, S. 95.
10. "Bahnstrecke Wadi Halfa-Kerma." Wikipedia, 2024, de.wikipedia.org/wiki/Bahnstrecke_Wadi_Halfa–Kerma.

Egypt
1. "Moving Egypt. Ein modernes Bahnsystem, das den Alltag neu erfindet", Siemens, 2022, press.siemens.com/global/de/feature/aegypten-bewegung.
2. Stephan Roll: "Kredite für den Präsidenten. Auslandsverschuldung und Herrschaftssicherung in Ägypten." SWP-Studie, 30. September 2022, doi.org/10.18449/2022S10v02.
3. Christiane Desroches-Noblecourt und Georg Gerster: "Die Welt rettet Abu Simbel." A. F. Koska, Wien-Berlin, 1968.
4. "The World Saves Abu Simbel." YouTube,

UNESCO Archives Film Collection, 1972, www.youtube.com/watch?v=OhtMtoex2uA.
5. Hassan Fathy: "Architecture for the Poor." The American University in Cairo Press, 1973.
6. "New Gourna Village." World Monuments Fund, 2021, www.wmf.org/project/new-gourna-village.
7. "Quarry." ACCM, 2018, www.accm.com.eg/about-us/quarry/.
8. "Working in Egypt's Limestone Quarries – in Pictures." The Guardian, 27. Februar 2020, www.theguardian.com/artanddesign/gallery/2020/feb/27/working-in-egypts-limestone-quarries-in-pictures.
9. "Naoum Shebib – The Architect." Naoum Shebib, www.naoumshebib.com/pages_en/architect_en.html.
10. Mourad Bakhoum: "Planning, Design and Construction Aspects of Rod El Farag Cable-Stayed Bridge over River Nile, Cairo, Egypt." IABSE Congress Nanjing, 2022, p. 2077.
11. Mourad Bakhoum: "A Brief History of Bridges in Egypt with Brief Discussion of Related Constructions." fib Bridge Engineering Conference, Egypt, 2000.
12. Michel Bakhoum: "Structural Mechanics." Structural Engineering Department, Cairo University, 2010.

Considerations

Universities

1. "Länderinformationen." DAAD, 2024, www.daad.de/de/laenderinformationen/.
2. "Bildungssystemanalyse individuell." DAAD, 2024, www.daad.de/de/bildungssystemanalyse-individuell/.
3. "Stellenbosch University." unibw, München, 2024, www.unibw.de/internationales/partnerhochschulen/kooperationen-nach-laendern/suedafrika/stellenbosch.
4. "Bauingenieurwesen Double Degree (NUST)." FH Aachen, 2024, www.fh-aachen.de/studium/studiengaenge/bauingenieurwesen-beng/studiengangsvariante/bauingenieurwesen-double-degree-nust.
5. "Neue Konzepte für El Gouna – Interview mit Präsident Prof. Dr. Christian Thomsen." TU Berlin: Referat Transnationale Bildung Und Internationaler Wissenstransfer, 2021, https://web.archive.org/web/20230321103727/www.tu.berlin/themen/internationales/neue-konzepte-fuer-el-gouna.
6. "All Countries Targeted for Zero-Carbon-Ready Codes for New Buildings by 2030." IEA, 2022, www.iea.org/reports/all-countries-targeted-for-zero-carbon-ready-codes-for-new-buildings-by-2030-2.
7. Udo Kultermann: "New Directions in African Architecture." George Braziller, New York, 1969.
8. Julia Watson: "Lo–TEK. Design by Radical Indigenism." Taschen, Köln, 2023.

Solar Energy

1. "Endenergieverbrauch nach Energieträgern und Sektoren." Umweltbundesamt, 2024, www.umweltbundesamt.de/daten/energie/energieverbrauch-nach-energietraegern-sektoren#allgemeine-entwicklung-und-einflussfaktoren.
2. "Erneuerbare Energien in Deutschland 2022: Daten zur Entwicklung im Jahr 2022." Umweltbundesamt, 2023, www.umweltbundesamt.de/publikationen/erneuerbare-energien-in-deutschland-2022.
3. "Africa Energy Outlook 2022." IEA, 2022, p. 17, 68, 101 and 185, www.iea.org/reports/africa-energy-outlook-2022.
4. Wolfgang Schiel und Jörg Schlaich: "Wie viel Wüste braucht ein Auto?" Energie im Wandel, Springer, Berlin / Heidelberg, 2001, doi.org/10.1007/978-3-642-56592-2_14 und www.sbp.de/app/uploads/2022/05/Wie_viel_Wueste_braucht_ein_Auto.pdf
5. "Kraftwerk Ouarzazate." Wikipedia, 2024, de.wikipedia.org/wiki/Kraftwerk_Ouarzazate.
6. "Marokko – EDF Konsortium gewinnt Ausschreibung ‚Noor Midelt I Solarpark'." Maghreb-Post, 22. Mai 2019, www.maghreb-post.de/marokko-edf-konsortium-gewinnt-ausschreibung-noor-midelt-solarpark/.
7. "List of solar thermal power stations." Wikipedia, 2024, en.wikipedia.org/wiki/List_of_solar_thermal_power_stations.
8. "List of photovoltaic power stations." Wikipedia, 2024, en.wikipedia.org/wiki/List_of_photovoltaic_power_stations.
9. David Ehl: "Ostafrika ist heiß auf Geothermie." DW, 24. Januar 2024, www.dw.com/de/ostafrika-ist-heiß-auf-geothermie/a-68062697.
10. "Kenya – Country Commercial Guide." ITA, 2022, www.trade.gov/country-commercial-guides/kenya-energy-electrical-power-systems.
11. "Solarbetriebene Zementwerke, Zementklinker mithilfe von Sonnenenergie." beton 9/2023, p. 326.
12. "Solar Sinter." Kayser Works, www.kayserworks.com/#/798817030644/.

Concrete

1. Chaojun Chu et al.: "Allelopathic Effects of Eucalyptus on Native and Introduced Tree Species." Forest Ecology and Management 323, 2014, p. 79, doi.org/10.1016/j.foreco.2014.03.004.
2. Siegbert Uhlig et al.: "Äthiopien. Geschichte, Kultur, Herausforderungen." Harrassowitz, Wiesbaden, 2018, p. 25.
3. World Bank Group, Forest Area (% of land area), https://data.worldbank.org/indicator/AG.LND.FRST.ZS.
4. Teshome Tefera Tola et al.: "Undertaking Urban Metabolism Analysis in a Data Poor Context: The Metabolism of Housing Construction Materials in Amdework – a Small Town in Ethiopia." International Journal of Urban Sustainable Development, 11.2, 2019, p. 172–188, doi.org/10.1080/19463138.2019.1618859.
5. "Unlocking Ethiopia's Urban Land and Housing Markets." World Bank, 2019, documents.worldbank.org/en/publication/documents-reports/documentdetail/549221572382742218/unlocking-ethiopias-urban-land-and-housing-markets-synthesis-report.
6. Gonche Girma und Tensaye Abate: "The Status of Wood Products Supply and Demand in Ethiopia: A Review." Journal of Economics and Sustainable Development 12.1, 2021, p. 17.
7. Werner Sobek: "non nobis – über das Bauen der Zukunft. Band 1." avedition, Stuttgart, 2022, p. 80.
8. "Zu fein – Wüstensand kein hochwertiger Baurohstoff." BGR, 2019, https://web.archive.org/web/20220607153928/https://www.bgr.bund.de/DE/Gemeinsames/Nachrichten/Aktuelles/2019/2019-08-06_wuestensand.html.
9. "Blähglas – Ökobilanz." Das Forum Nachhaltiges Bauen, 2024, nachhaltiges-bauen.de/baustoffe/Blähglas.
10. "Dekarbonisierung von Zement und Beton – Minderungspfade und Handlungsstrategien.", VDZ, Düsseldorf, 2020, p. 11 and 35.
11. Mike Schlaich, Alexander Hückler und Claudia Lösch: "Infraleichtbeton (ILC)." Beton-Kalender, Ernst & Sohn, Berlin, 2021, p. 907–952.
12. "Mineral Exploration and Reserve Estimation Directorate." Amhara National Regional State Bureau of Mines Resource Development, 2024.
13. "Featured Products." Derba MIDROC Cement, 2018, https://web.archive.org/web/20240529182841/https://www.derbacement.com/products/.

Bridges

1. Graham Ross: "The Romance of the Cape Mountain Passes." Sunbird Publishers, 2011, S. 73.
2. "List of bridges in South Africa." Wikipedia, 2024, en.wikipedia.org/wiki/List_of_bridges_in_South_Africa.
3. Mark Thomas und Henry Fagan: "The Boomslang: Kirstenbosch Centenary, Tree Canopy Walkway." Sarah Thomas, Cape Town, 2014.
4. John Anderson et al.: "Designing for the Realities of a Day in the Life of a Good Looking but Hardy Footbridge." Footbridge 2017 Berlin – Tell a Story, TU Berlin, 2017, p. 80.
5. Mourad Bakhoum: "A Brief History of Bridges in Egypt with Brief Discussion of Related Constructions." fib Bridge Engineering Conference, Egypt, 2000.
6. "Category: Cable-stayed bridges in Egypt." Wikipedia, 2020, en.wikipedia.org/wiki/Category:Cable-stayed_bridges_in_Egypt.
7. "Structurae: Internationale Datenbank und Galerie für Ingenieurbauwerke." Structurae, 2024, www.structurae.net/de.
8. Chris Rollins: "Lessons from the Field." Footbridge 2005, Venezia 2005, S. 189.
9. "Bridges to Properity." 2024, https://web.archive.org/web/20260105120240/www.bridgestoprosperity.org/.
10. "Liste der Rheinbrücken." Wikipedia, 2024, de.wikipedia.org/wiki/Liste_der_Rheinbrücken.

Roads

1. Felix Guntermann: "Verkehrsinfrastruktur in Afrika. Anbindung als Aufgabe und Chance." Africa Business Guide, 10. August 2020, www.africa-business-guide.de/de/chancen/verkehrsinfrastruktur-in-afrika-537300.
2. "Tracking Africa's Progress in Figures, Chapter 5: Infrastructure Development." African Development Bank, 2014, S. 51.
3. "Rural Access Index (RAI)." World Bank Data Catalog, 2024, datacatalog. https://unstats.un.org/sdgs/metadata/files/Metadata-09-01-01.pdf.
4. Ryszard Kapuscinski: "Ein Loch in Onitsha." In: Afrikanisches Fieber. Erfahrungen aus vierzig Jahren, Malik National Geographic, 2019.
5. "Africa Status Report on Road Safety 2020." SSATP, 2021, www.ssatp.org/publication/africa-status-report-road-safety-2020.
6. "How New Road Systems Support Egypt's Economy." OBG, https://web.archive.org/web/20250615191131/oxfordbusinessgroup.com/reports/egypt/2020-report/economy/key-connections-government-increases-investment-in-road-infrastructure-to-ease-congestion-and-support-economic-development.
7. "Trans-African Highway network." Wikipedia, 2024, en.wikipedia.org/wiki/Trans-African_Highway_network.
8. "Siemens Mobility schließt Vertrag über 2.000 km langes Hochgeschwindigkeitssystem in Ägypten ab." Siemens, 28. Mai 2022, press.siemens.com/global/de/pressemitteilung/siemens-mobility-schliesst-vertrag-ueber-2000-km-langes-hochgeschwindigkeitssystem.
9. Malcolm Mitchell und Brian Downie: "Travels in the South – A History of South African Roads." Media24 famous publishing, Durban, 2018.

China

1. David H. Shinn: "China–Africa Ties in Historical Context." China-Africa and an Economic Transformation, Oxford University Press, April 2019, S. 61–83, doi.org/10.1093/oso/9780198830504.003.0004.
2. "China-Africa Data." China Africa Research Initiative at Johns Hopkins University's School of Advanced International Studies (CARI), 2024, www.sais-cari.org/data.
3. "Bauindustrie in Afrika." Hauptverband der Deutschen Bauindustrie, 2018, www.bauindustrie.de/fileadmin/bauindustrie.de/Media/Veroeffentlichungen/Bauindustrie-Positionspapier-Afrika-WEB-final.pdf.
4. Hong Zhang: "Chinese International Contractors in Africa: Structure and Agency." CARI, working paper 2021/47, 2021, www.econstor.eu/bitstream/10419/248175/1/sais-cari-wp47.pdf.
5. Howard W. French: "China's Second Continent: How a Million Migrants are Building a New Empire in Africa." Vintage Books, New York, 2014.
6. Simon Züfle: "Die Seidenstraßeninitiative in Ostafrika: Auf dem Weg zu einer Hegemonie Chinas?" Springer VS, Wiesbaden, 2022, p. 192 and 295.
7. Jing Gu und Richard Carey: "China's Development Finance and African Infrastructure Development." China-Africa and an Economic Transformation, Oxford University Press, 2019, doi.org/10.1093/oso/9780198830504.003.0008.
8. Deborah Brautigam und Won Kidane: "China, Africa and Debt Distress: Fact and Fiction about Asset Seizures." CARI, policy brief No. 47/2020, 2020, www.econstor.eu/handle/10419/248226.
9. Bénédicte Savoy: "Afrikas Kampf um seine Kunst. Geschichte einer postkolonialen Niederlage." C.H. Beck, München, 2021.

Residential Construction

1. "World Population Prospects 2024." United Nations, Department of Economic and Social Affairs, Population Division, population.un.org/wpp/Graphs/DemographicProfiles/Line/903.
2. Bruno Urmersbach: "Afrika: Die 20 größten Städte im Jahr 2023." Statista, 2024, de.statista.com/statistik/daten/studie/946810/umfrage/groesste-staedte-in-afrika/.
3. Andres Lepik: "Afritecture." Hatje Cantz, Ostfildern, 2013.
4. "Eine Reise zum nachhaltigen Bauen." OSTpunkt University Magazine, 2023, www.issuu.com/ostschweizerfachhochschule/docs/ostpunkt-inhalt_1-2023-web_300/s/19518403.
5. Fabienne Hoelzel: "Lagos, Nigeria: A Study in Anarchy and Negotiation." Bhúmi, The Planning Research Journal 9.1, 2022, p. 18–34, doi.org/10.4038/bhumi.v9i1.86.
6. "Oworonshoki – Building Guide for a Flood Resilient Waterfront Community." Fabulous Urban – Nigeria Foundation, 2023, www.academia.edu/111930450/OWORONSHOKI_BUILDING_GUIDE_for_a_Flood_Resilient_Waterfront_Community.
7. Katharina Lux: "Postkoloniale Stadtplanung in der Lehre: Ein paar Fragen an Fabienne Hoelzel." Baunetz, 17. Oktober 2023, www.baunetz-campus.de/people/postkoloniale-stadtplanung-in-der-lehre-ein-paar-fragen-an-fabienne-hoelzel-8370522.
8. Luyanda Mpahlwa: "Reclaiming the Past: Our Heritage, Our Cultures, Our Future." Architectural Guide Sub-Saharan Africa, Volume 1, DOM publishers, Berlin, 2021, p. 178.
9. "African Centre for Cities – Urbanism from an African Perspective." University of Cape Town, 2024, www.africancentreforcities.net.

All websites were last accessed on 12 July 2024.

The German National Library lists this publication in the German National Bibliography; detailed bibliographic data is available on the Internet at http://dnb.d-nb.de.

ISBN 978-3-86922-933-1 (softcover)
ISBN 978-3-86922-970-6 (hardcover)

www.dom-publishers.com

Photos
Mike Schlaich

Editing
Ursula Baus, Uta Keil

Proofreading
Sandie Kestell

Design
Moniteurs, Berlin
Sibylle Schlaich, Annick Rietz, Anna Gaißmaier

Printing
Druckhaus Sportflieger, Berlin

Paper
Arctic Volume White, 130 g/m²

Fonts
Euclid Flex, Trump Mediaeval

DOM
publishers